THE ORIGIN AND EVOLUTION OF HUMANS AND HUMANNESS

EDITED BY

D. TAB RASMUSSEN

JONES AND BARTLETT PUBLISHER

Boston   **London**

Editorial, Sales, and Customer Service Offices

Jones and Bartlett Publishers
One Exeter Plaza
Boston, MA 02116
1-617-859-3900
1-800-832-0034

Jones and Bartlett Publishers International
P O Box 1498
London W6 7RS
England

Copyright © 1993 by Jones and Bartlett Publishers, Inc.

All rights reserved. No part of the material protected by this copyright notice may be reproduced or utilized in any form, electronic or mechanical, including photocopying, recording, or by any information storage and retrieval system, without written permission from the copyright owner.

Library of Congress Cataloging-in-Publication Data

The Origin and evolution of humans and humanness / edited by D. Tab
  Rasmussen.
     p.   cm.
  "A contribution of the IGPP Center for the Study of Evolution and
the Origin of Life (CSEOL), University of California, Los Angeles"-
-T.p. verso.
  Includes bibliographical references and index.
  ISBN: 0-86720-857-0
  1. Human evolution. I. Rasmussen, D. Tab. II. University of
California, Los Angeles. IGPP Center for the Study of Evolution and
the Origin of Life.
GN281.0736 1993
573.2—dc20
                                            93-6741
                                               CIP

Production Editor: Joni Hopkins McDonald
Editorial Production Service: The Book Company
Cover Design: Hannus Design Associates
Printing and Binding: Courier Westford

A contribution of the IGPP Center for the Study of Evolution and the Origin of Life (CSEOL), University of California, Los Angeles.

Printed in the United States of America
97 96 95 94 93   10 9 8 7 6 5 4 3 2

CONTENTS

CHAPTER 3   **ARCHEOLOGICAL INTERPRETATIONS OF EARLY HOMINID BEHAVIOR AND ECOLOGY   49**

Richard Potts

PREFACE

On March 13, 1992, the Irving Stone Memorial Symposium on "The Origin and Evolution of Humans and Humanness" was held on the campus of the University of California, Los Angeles, attended by hundreds of students and other members of the Los Angeles community. The symposium was convened by the Center for the Study of Evolution and the Origin of Life (CSEOL), a group of faculty, researchers, graduate students and other academics who, when they want to know the state of the art about something, invite the experts to tell them. Accordingly, six world-class scientists covering the fields of anthropology, archeology, biology and ecology were brought to the UCLA campus to report their latest findings and to express their interpretations concerning the evolutionary history of human beings. This volume makes accessible the proceedings of that symposium.

The symposium participants were invited not only for their ability to cover the full time-depth of human evolution, from our apelike ancestors of 10 million years ago up to the biological and cultural processes operating in modern humans today, but also for the spectrum of disciplines and variety of approaches they bring to bear on the questions of how, where, and why humans evolved. Chapter 1 describes what made the earliest humanlike animals 4 million years ago different from their ape relatives, drawing heavily on knowledge of the behavior and reproduction of living primates. Chapter 2 focuses on the origin of our own genus, *Homo,* while also demonstrating how the science of paleontology works. Chapter 3 uses an archeological approach to interpreting ancient human behavior and ecology, especially focusing on those human forebears who first made regular use of stone tools 2 million years ago. Chapter 4 examines the origin of our own species—modern *Homo sapiens*—by synthesizing knowledge gained in genetics, molecular biology, archeology, and paleontology. Chapter 5 describes the beautiful richness and diversity of Ice Age imagery, while also exploring the symbolic behavior of ancient artists and craftsmen. Finally, Chapter 6 addresses some of the most difficult questions of all—how does human behavior change?, and what is the relationship between biological and cultural evolution?—using computer methods to help us grasp the complex but precise logic of evolutionary change.

This book is intended for a broad audience of interested scientists, naturalists, and philosophers, but it is especially designed for use in college coursework on human evolution. The six authors provide an up-to-date review of new fossil and archeological evidence, but they also share their expert interpretations of the larger questions concerning the emergence of humanness. For easy reference, a time chart of human evolution appears on the inside front cover, and a diagram of ape and human skeletal anatomy is provided on the inside back cover. A glossary of technical terms is included after the last chapter; glossary entries appear in boldface type where they are first encountered in the text. Biographical sketches of the authors follow, along with a tribute to Irving Stone, the man who inspired this symposium.

D. Tab Rasmussen
Washington University
St. Louis

IRVING STONE

Irving Stone *1903–1989*

This volume, like the symposium from which it was derived, is dedicated to Irving Stone, founding father of the biographical novel and a long-time member of CSEOL. The works of Irving Stone are legendary—author of more than 15 major novels that bring to life such figures as Vincent Van Gogh (*Lust for Life,* 1934), Jack London (*Sailor on Horseback,* 1938), Michelangelo (*The Agony and the Ecstasy,* 1962), and Sigmund Freud (*Passions of the Mind,* 1971). Irving was, in fact, "a legend in his own time." But to his professional friends and colleagues in CSEOL, it was his Darwin book (*The Origin,* 1980) that made him one of us. Here was a subject that we evolution-type academics knew we knew. And here was a writer who knew far more than did we. He beat us at our own game. He was wonderful!

To CSEOL, however, Irving was much more than "merely" our star Writer in Residence. He was an inspiration, a reservoir of wisdom, and a scholar to the core. He understood knowledge and its intrinsic value:

> All ignorance is bad and all knowledge good.
> We were born into this world a long time ago.
> In the beginning we knew nothing of the forces surrounding us.
> But for all these millions of years the human brain has been chipping away
>   at that ignorance, storing up hard earned wisdom.
> This is the greatest adventure of mankind: *To find something that was never*
>   *known before, or understood.*
> Each new piece of knowledge does not need to have a specific or functional
>   use, at least not at the moment.
> It is sufficient triumph that we have learned something that had formerly
>   been part of the darkness.

And Irving Stone was an Idea Man. He was fascinated by the yet unsolved, especially if it pertained to the human condition. One of his favored questions, the notion that led to this symposium, was "When did man become Man? . . . What are the evolutionary roots of humanness?" Irving suggested, indeed urged, that CSEOL examine this issue. His idea. His symposium. He was the author. The symposium participants have pursued that quest.

J. William Schopf, Director
Center for the Study of Evolution
   and the Origin of Life
University of California
Los Angeles

BIOGRAPHICAL SKETCHES

Chapter 1: Modeling Human Origins: Are We Sexy Because We're Smart, or Smart Because We're Sexy?

C. OWEN LOVEJOY

C. Owen Lovejoy
(Kent State University)

Dr. Lovejoy received his Ph.D. from the University of Massachusetts in 1970; he is currently Professor of Anthropology at Kent State University. He is a prolific author with a wide range of research talents in the anatomical, behavioral, and biomolecular sciences, with special interests in behavioral and functional correlates of osteology (the study of bones). Dr. Lovejoy has devoted particular attention to mammalian locomotor adaptations and is recognized as the world's foremost authority on the unique human attribute of upright, bipedal walking. His analyses and interpretation of the skeleton of the famous "Lucy" and her kin—small, bipedal human ancestors dating from more than 3 million years ago—have proven instrumental in our understanding of hominid origins.

Chapter 2: The Origin of the Genus *Homo*

ALAN WALKER

Alan Walker
(Johns Hopkins University)

Born in Leicester, England, and educated at Cambridge University (B.A., 1962) and at the University of London (Ph.D., 1967), Dr. Walker taught at Makere University in Uganda, the University of Nairobi, and at Harvard University before assuming his present position of Professor in the Department of Cell Biology and Anatomy at Johns Hopkins University. He has served as Director of the Foundation for Research into the Origin of Man, and is the recipient of many honors and awards including Guggenheim and MacArthur Foundation Fellowships, and the CSEOL Distinguished Scientist Award. Dr. Walker's research in primate evolution has spanned a broad spectrum of anatomical, behavioral, and paleontological studies, but he is perhaps best known for his seminal investigations of early hominids from Kenya.

Chapter 3: Archeological Interpretations of Early Hominid Behavior and Ecology
RICHARD POTTS

Richard Potts
(Smithsonian Institution)

Born in Philadelphia, Dr. Potts was educated at Temple University (B.A.) and at Harvard University where he received his Ph.D. in 1981. He currently serves as Associate Curator of Anthropology at the National Museum of Natural History, Smithsonian Institution. He is an acknowledged authority on the paleoecology and behavior of early hominids, a reputation stemming from his original, creative fieldwork, and his talent for synthesizing ecological theory and archeological methods, as reflected in his influential book *Early Hominid Activities at Olduvai*. Renowned for his research at Olduvai Gorge and other East African early hominid sites, Dr. Potts is now expanding the scope of his efforts by investigating paleoecological and archeological aspects of human evolution in China and Java.

Chapter 4: New Views on Modern Human Origins
CHRISTOPHER B. STRINGER

Christopher B. Stringer
(The Natural History Museum, London)

A British citizen, Dr. Stringer earned his B.Sc. from University College, London (1969) and his Ph.D. from the University of Bristol (1974). During 1979, he was a Visiting Lecturer in the Department of Anthropology, Harvard University; currently he is Head of the Human Origins Programme at The Natural History Museum, London. Dr. Stringer's investigations into the fate of Neanderthals and of the origin of anatomically modern humans have drawn eclectically from anatomical, biomolecular, and paleontological evidence. His own highly significant research contributions, together with his exceptional ability to synthesize data from disparate fields of the science, have made him a pivotal figure in the "Out of Africa" theory of modern human origins.

Chapter 5: Humans as Materialists and Symbolists: Image Making in the Upper Paleolithic
MARGARET W. CONKEY

Margaret W. Conkey
(University of California, Berkeley)

Educated at Mount Holyoke College (B.A.) and the University of Chicago (Ph.D., 1978), Dr. Conkey taught at the University of California, Santa Cruz, and at the State University of New York, Binghamton, before moving to her current position as a faculty member at the University of California, Berkeley. She is an authority on the art and visual imagery of Ice Age Europe, conducting regular archeological fieldwork in the French Pyrenees. Dr. Conkey's research focuses on how Ice Age peoples made use of available natural materials for the technical production of images, and what this material aspect of the images may reveal about their symbolic meaning, or about the behavior of the image makers. She is also an internationally recognized pioneer in assessing issues of gender in archeology.

Chapter 6: Culture and Human Evolution
ROBERT BOYD and PETER J. RICHERSON

Robert Boyd
(University of California,
Los Angeles)

Dr. Boyd, who received his B.A. in Physics from the University of California, San Diego, and his Ph.D. in Ecology from U.C., Davis, is now Professor of Anthropology at UCLA. Dr. Richerson is Professor in the Institute of Ecology, University of California, Davis. Drs. Boyd and Richerson, both of whom specialize in population genetics and evolutionary mechanisms, began their productive collaboration while Boyd was a student of Richerson's at UC Davis. Together, they bring a broad, interdisciplinary perspective and technical sophistication to their studies of the complex phenomenon of cultural evolution. Their book *Culture and the Evolutionary Process* won the 1989 J. I. Staley Prize awarded by the School of American Research; this work and others have earned them international recognition as leaders in the study of human behavior and adaptation.

MODELING HUMAN ORIGINS: ARE WE SEXY BECAUSE WE'RE SMART, OR SMART BECAUSE WE'RE SEXY?

■

C. Owen Lovejoy*

■

MODELING HUMAN EVOLUTION: THE SCENARIO

At the beginning of a criminal trial, the prosecutor addresses the jury with an opening statement in which he or she presents a "theory of the crime." This theory is usually a hypothetical narrative weaving together the people, motives, events, and physical facts of the case before the bar. This "scenario" must be clear, well-organized, and persuasive because the witnesses who will follow can only testify to isolated facts. If no one actually witnessed the crime, all evidence is circumstantial and the jury will have to make its decision solely on the weight of those facts.

There are many parallels between a prosecutor's theory and a biologist's attempt to reconstruct evolution from the fossil record. There are obviously no witnesses to past evolution, only a trail of evidence that must be woven into an *evolutionary* scenario. In a prosecutor's theory there is almost always a single organizing theme that serves as its engine—the *motive* of the crime. In an evolutionary scenario there is an equally important and compelling engine—**natural selection**—which must also be used to fuse the various elements into a cohesive whole. The prosecutor's narrative may lack many details of the crime and contain many gaps—he or she did not actually witness it and is only reconstructing its details from isolated facts. As evolutionary biologists, there are many aspects of our own evolution about which we will never know simply because the evidence is incomplete. And yet, just as for the prosecutor, there is every reason to believe that we can also reach a verdict, that we can solve the fundamental mysteries of our own origins.

Human evolution is one of the most intriguing of all evolutionary puzzles. When and where did our species acquire its peculiar form of locomotion, massive brain, and ability to communicate with complex symbols? We would have none of these characteristics today had our most ancient ancestors not embarked on an evolutionary journey distinct and separate from that taken by our closest living relatives, the great apes. In many ways our ancestors had to set the stage for the further develop-

*Department of Anthropology, Kent State University, Kent, OH 44242

ment of these characteristics, even though they themselves never benefited from some of them (such as language or technology). In this chapter we explore some of the anatomical facts with which an evolutionary model can be constructed and attempt to weave them into a cohesive scenario.

How can the probable accuracy of such a model be judged? Evolutionary scenarios must be evaluated much in the same way that jury members must judge a prosecutor's narrative. Ultimately they must make their judgment not on the basis of any single fact or observation, but on the totality of the available evidence. Rarely will any single item of evidence prove pivotal in determining whether a prosecutor's scenario or the defense's alternative is most likely to be correct. Many single details may actually fail to favor one scenario over another. The most probable account, instead, is the one which is the most internally consistent—the one in which all the facts mesh together most neatly with one another and with the motives in the case. Of paramount importance is the *economy of explanation*. There are always alternative explanations of any single, isolated fact. The greater the number of special explanations required in a narrative, however, the less probable its accuracy. An effective scenario almost always has a compelling facility to explain a chain of facts with a minimum of such special explanations. Instead the pieces of the puzzle should fall into place.

THE SETTING OF HUMAN EVOLUTION

The Emergence of the Hominoids of the Miocene

Our species, *Homo sapiens*, is the only living member of a family of primates called the **Hominidae**. Together with the living great apes (orangutan, gorilla, chimpanzee, and **bonobo**), we are members of a larger taxonomic group (usually recognized at the superfamily level) generally called the **Hominoidea (hominoids)**. The taxonomic separation of humans from the other living apes is more traditional than biologically justified. In fact, humans are remarkably similar to the living apes in much of their basic structural anatomy and physiology, and are strikingly similar to them when distance is judged on purely genetic grounds. Many aspects of our **karyotypes** (the number and structure of our **chromosomes**) show virtual identity with other apes, and hybridization of our respective **DNA** indicates that over 98% of the **nucleotide base** pairs in all of our nonrepeated DNA are identical (Sibley and Ahlquist, 1987). Current biochemical studies of our **genomes** indicate that we are approximately equal in the degree to which we differ from each of the African hominoids (bonobo, chimpanzee and gorilla), but significantly more distant from the only living Asian hominoid, the orangutan (Goodman et al., 1990). Chimpanzees, bonobos, gorillas, and humans therefore constitute a closely circumscribed evolutionary group.

Hominoids made their first appearance in the early **Miocene** almost 25 million years ago in East Africa (Figure 1.1; Kelley and Pilbeam, 1986). At that time they differed from their descendants in a variety of important ways and might strike the reader as being somewhat more like monkeys than apes in their postures and locomotion. We can recognize them as our linear antecedents, however, because they display many detailed traits of dental anatomy that are shared by all hominoids. In the period between their first appearance and about 8 to 10 million years ago, fossil hominoids can be seen to undergo progressive change in the direction of their living descendants, and several recent finds shed considerable new light on their evolution.

One of the most dramatic is a specimen from Pakistan known simply as GSP-

FIGURE 1.1

Approximate distribution of the Tethys Sea (current remains are the Mediterranean Sea, Red Sea, and Persian Gulf) and the Paratethys Sea (current remnants include the Caspian and Black Seas) during the early Miocene. During this period the Afro-Arabian Plate was isolated from Eurasia by these two extinct seas. The probable migration routes used by hominoids after docking of the Afro-Arabian plate with Eurasia (at about 18 to 16 million years) are indicated. Redrawn after Brown, B., *The Mandibles of Sivapithecus*, Ph.D. dissertation, Kent State University, Kent, Ohio (1989).

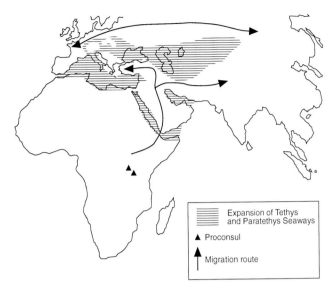

15000,* which is assigned to the genus **Sivapithecus** (Figure 1.2). It has an approximate age of 8 to 9 million years. Its most striking feature is the degree to which its facial skeleton adumbrates that of our living "cousin"—the modern orangutan (Ward and Brown, 1986). This remarkable similarity implies that apes had become relatively modern by the time of their appearance in the fossil record.

During the earliest hominoid period (between 25 and 17 million years ago) Africa and Arabia were separated from the rest of Eurasia by shallows seas that at times connected the Atlantic and Indian Oceans, isolating Africa as an immense island continent (Figure 1.1). Ancestral hominoids were restricted to this island until about 17 million years ago, when Africa docked, so to speak, with Eurasia (Steininger et al., 1985). This event opened a passageway by which mammals could enter and leave Africa, and it is clear from the fossil record that hominoids were exceptionally capable of doing so. They were highly successful, radiated widely, and were more prolific than all other primates in the post-Tethys period (Bernor, 1983).

Another familiar group of primates, the **Old World monkeys,** were far less successful during this same time period. They were less abundant and less diverse than the contemporary hominoids. Inspection of the fossil record reveals, however, that over the course of the last 10 million years, Old World monkeys have almost completely *replaced* hominoids as the dominant Old World primate (Figure 1.3). Why did this wholesale replacement occur? It might at first seem irrelevant to the problem originally posed at the beginning of this chapter—that is, the origin of our human ancestors. To the contrary, it provides a fundamental clue: while our extremely close relatives, the great apes, were being almost totally replaced by the radiating Old

---

*Only occasionally do individual fossil specimens earn their own name (for example, the famous "Lucy" skeleton). Usually, fossils are identified by museum numbers which facilitate record keeping and allow precise communication among researchers. Museum numbering systems that appear in this chapter identify fossils from Ethiopia and Pakistan. Specimens from Pakistan bear the prefix GSP (for Geological Survey of Pakistan) and then are numbered sequentially in the approximate order of discovery. Specimens from the Afar Triangle in northern Ethiopia bear the prefix AL (for Afar Locality), followed by the site number and then the specimen number (AL-288-1 is the first specimen found at site #288 in the Afar Triangle).

FIGURE 1.2

Direct lateral and three-quarter views of GSP-15000 (center) compared to a modern orangutan (right) and a modern chimpanzee (left). The striking resemblance of the fossil with the orang is evident. Note the narrow interorbital region, the similar conformation of the supraorbital regions, the similarity in incisor position and relative size, the vertically elliptical orbits, the conformations of the cheek areas, and the conformation and positioning of the canine. Photo by D. Brill.

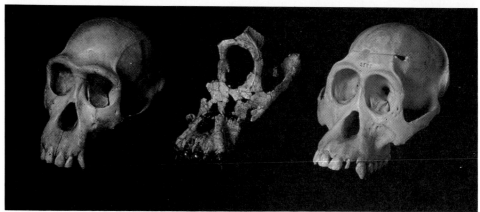

FIGURE 1.3

Species diversity in hominoids and **cercopithecoids** during the last 25 million years. Percent of total higher primate species that are hominoid or cercopithecoid is shown for four dates of assessment. Note the inverse relationship between these two taxa during the last 25 million years. Redrawn after Andrews, 1981.

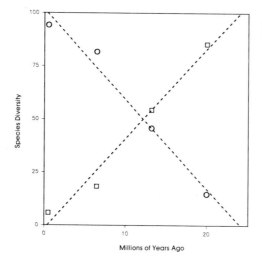

World monkeys, one group of apes—human ancestors—continued to thrive. What special adaptive ability did they possess that was not shared by their closely related contemporaries?

■
THE AUSTRALOPITHECINES

The Earliest Fossil Hominids

We do not know when our oldest direct ancestors first separated from the African apes. The event almost certainly took place sometime after the ancestors of the Asian apes had migrated out of Africa. As noted earlier, orangutans differ from humans and the African apes about twice as much as these hominoids differ from one another. This means that our last common ancestor with the bonobo, chimpanzee, and gorilla should be about one half as ancient as that which we all share with the orangutan. Therefore, if the ancestors of the modern orangutan did leave Africa at about the time that the Afro-Arabian **continental plate** docked with the remainder of Eurasia (about 17 million years ago), then our last common ancestor with the African apes would be approximately 8 to 9 million years old.

Unfortunately, no fossils have yet been recovered from this time period that could confirm these estimates. Indeed, the first recognizable **hominids**, an extinct group known as the **australopithecines**, have been found no older than about 4 million years. Although some fragmentary remains of australopithecines have been recovered from sites as early as 5.5 million years, the earliest specimens that can be used to reconstruct the details of their biology are from East African sites less than 4 million years old (Figure 1.4; Hill and Ward, 1987).

Australopithecines shared many features of their dental, facial, and locomotor anatomy with the African apes (Figure 1.5). It is, however, the ways in which they had already evolved distinct differences from apes that promise the greatest insight into their first appearance and evolution. We concentrate on three of the most basic differences because these provide the most telling evidence about the fundamental biological nature of these ancient human ancestors.

The Unique Dental Features of the Australopithecines

In its overall structure, the dentition of the earliest known australopithecine **species**, *Australopithecus afarensis*, is quite similar to that of our living ape relatives. The upper jaw and face in this species are strikingly like those of a modern chimpanzee (Figure 1.5). A closer look, however, reveals two critically important differences. First, the **molar** teeth display a variety of changes that indicate a different diet from a chimpanzee's. Compared to the molars of other African apes, those of early hominids have much thicker **enamel** and lower, stouter crowns with more rounded individual cusps (Johanson et al., 1982; Picq, 1991).

Chimpanzees are largely **frugivorous** and gorillas are mostly **folivorous**. Both diets require a dentition in which plant cells are sheared between upper and lower cusps. This action breaks up plant cells so that they release their contained nutrients. Such shearing is most effectively produced by thin enameled, relatively high-crowned teeth. By contrast, the molars of the earliest australopithecines are much blunter and flatter—less effective for shearing foodstuffs, but much more capable of resisting

FIGURE 1.4

Map of East Africa showing major fossil sites.

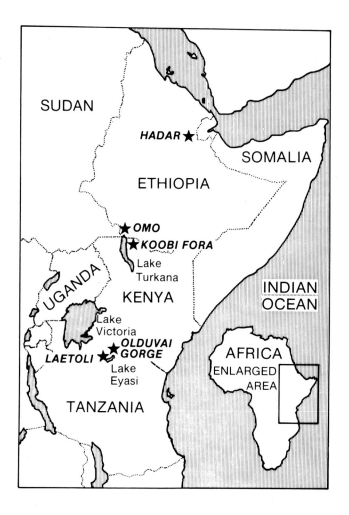

extreme wear and other damage caused by hard dietary inclusions. A simple and direct way to understand the purpose of such adaptations is to compare australopithecine molars to those of other mammals. Interestingly, terrestrial **omnivores** such as some species of bears and pigs, have molar crown and enamel structure quite similar to those of early hominids (Figure 1.6; Hatley and Kappelman, 1980).

A second feature that distinguishes the dentitions of australopithecines from those of apes is the degree of **canine dimorphism**. Unlike those of other hominoids, the canines of *A. afarensis* (and all later australopithecine species) are small and are of roughly similar size in both sexes (Johanson and White, 1979). Two possibilities exist for this reduction of canine dimorphism in our earliest ancestors. First, in a few rare cases, canines have been reduced in some primates as part of a specialized dietary adaptation. In these species, however, there are accompanying changes in the **incisor** teeth that signal the functional nature of this specialization. In the earliest australopithecines, the incisors remain largely unchanged from their size and form in other hominoids, greatly reducing the probability that this is an accurate explanation of canine reduction.

A much more probable explanation is a non-dietary one. Most primates have **polygynous** social structures in which males engage in a variety of intrasexual aggressive behaviors, in which canines play a key role (Figure 1.7; Jolly, 1985). In monogamous species such as gibbons, however, the canines of the two sexes differ very little (Figure 1.8). In many of these, such canine **monomorphism** has resulted from enlargement of the female canine rather than reduction of the male's. Many

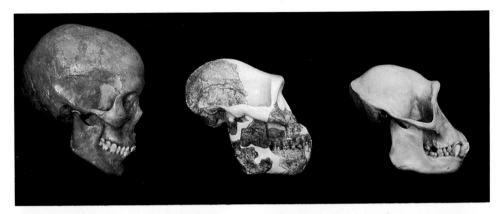

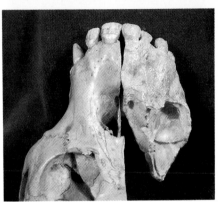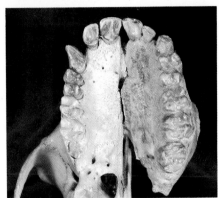

FIGURE 1.5

The cranium and facial structure of *Australopithecus afarensis*. In the top view, a reconstruction of the skull of this early hominid (see White et al., 1981) is shown compared to a modern human (left) and a modern chimpanzee (right). Note the overall similarity of the early hominid and chimpanzee. The early hominid demonstrates neither the striking brain expansion nor the extreme dental reduction seen in the modern human. In the lower two views, a maxilla (upper jaw) from *A. afarensis* that was naturally broken along its midline is shown articulated with a chimpanzee cranium that has been sectioned through its midline. Note the remarkable overall similarity in facial and dental structure in these two specimens. The primary differences are in canine size (the australopithecine specimen shown is a probable male) and in molar size. The australopithecine specimen's molars are megadontic relative to those of the chimpanzee; in addition, its molar crowns are lower and its enamel is thicker than those of the chimpanzee.

monogamous species live as isolated pairs who defend their territory from exploitation by other **conspecifics.** The female helps defend the pair's territory and has evolved the equipment necessary to accomplish this task.

The virtual absence of canine dimorphism in australopithecines therefore strongly implies that they may have been monogamous. But can we be sure that their small male canines were not the result of an unusual dietary adaptation? There is one additional but important clue. If the reduction of the canine were a response to dietary selection, we would expect the entire tooth to have been reduced. The male canines of the earliest australopithecines, however, show differential reduction of the canine crown, while the roots of these teeth continue to show distinct sexual dimorphism (Figure 1.9). This would suggest a selective response to social behavior, rather than diet. But what social selective force could favor males with smaller canine crowns?

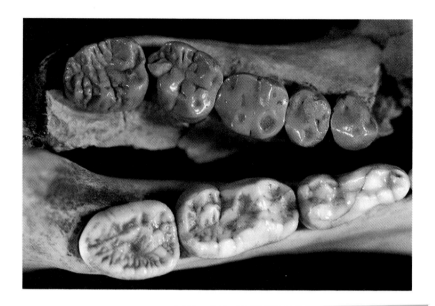

FIGURE 1.6

Comparison of the mandibular cheek teeth of an australopithecine (AL-288-l, otherwise known as "Lucy") and those of a North American black bear (*Ursus arctos*). Only the molars of the bear are shown; the premolars of the australopithecine specimen are also included. Note the thickened enamel, low-crown relief, and overall similarity of conformation in these two totally unrelated species. Note also that the second molar of the bear has the same approximate crown area as the combined first and second molar in the australopithecine, and that the australopithecine premolars are generally equivalent in size and morphology to the first molar of the bear. The premolars of bears are not chewing teeth (they are exceptionally small and peglike), so that the above is a direct comparison of the crushing and mastication roles of the dentitions in the two species. For further discussion see Hatley and Kappelman, 1980.

FIGURE 1.7

Male (left) and female (right) skulls of *Theropithecus gelada* (the gelada baboon). Note the massive canine in the male specimen. This is an extreme example of canine dimorphism in primates.

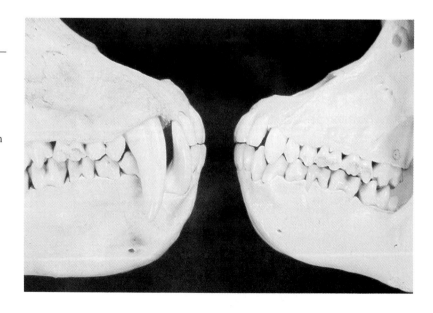

FIGURE 1.8

Male (left) and female (right) skulls of *Hylobates lar* (the gibbon). Note the overall similarity in size of the canines (canine monomorphism) achieved by enlargement of the female canine. Compare to Figure 1.7.

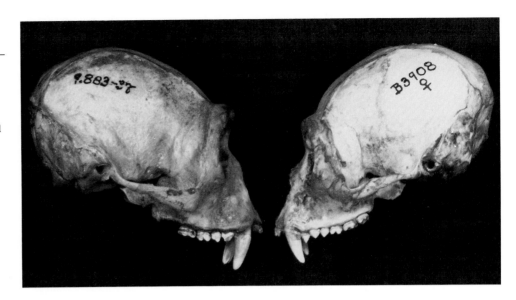

As we have just seen, most monogamous primates obtain canine monomorphism by enlarging the female canine, not reducing that of the male. We discuss this important clue in a following section.

The Unique Locomotor Anatomy of the Australopithecines

Another vitally important character by which *A. afarensis* differed from previous hominoids was its locomotor pattern. We had to look at the face and dentition in some detail to isolate particular differences between australopithecines and the African apes, but such a fine-grained analysis is not necessary when we compare their **postcranial** skeletons—here the differences are both multiple and striking.

Although a large number of individual fossil specimens have been attributed to *A. afarensis*, the best known is a partial skeleton labeled AL-288-1, often referred to as "Lucy" from the site of Hadar, Ethiopia (Figure 1.10). Hadar has also yielded an articulated foot (AL-333-115), a beautifully preserved knee joint (AL-128), and numerous other fossils that can guide our understanding of their locomotor pattern (Johanson et al., 1982). Although the Hadar skeleton still retains some anatomical details that are apelike (especially in forelimb structures such as the hand), the lower limb demonstrates that hominids had been walking **bipedally** for a long time because it is remarkably similar to that of modern humans. Their **pelvis** had undergone a complete reorganization, making it shorter, broader, and more fully suited to perform the tasks required of a biped (Figure 1.11; Lovejoy, 1988). The knee joint dem-

FIGURE 1.9

A male canine from *A. afarensis* (AL 333X-3). Note the moderate crown size compared to the very massive root. For discussion see text.

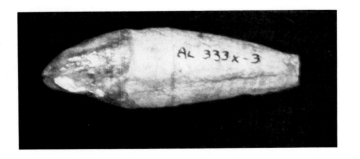

FIGURE 1.10

Full skeletal reconstruction of AL
288-1 ("Lucy"). The darkened
areas of each skeletal element are
reconstructed, although a large
proportion of this reconstruction
is based simply on bilateral
symmetry. The pelvis, femur,
tibia, ulna, radius, and humerus
have virtually complete skeletal
morphology and required only
shaft interpolation and estimation
of overall length. The skull is
largely conjectural and is one
suggestion of a female
counterpart to the male skull
shown in Figure 1.5. The clavicle,
foot, hand, and spine are based
on other (non-"Lucy") specimens
from Hadar and the footprints
from Laetoli, Tanzania. Photo by
D. Brill.

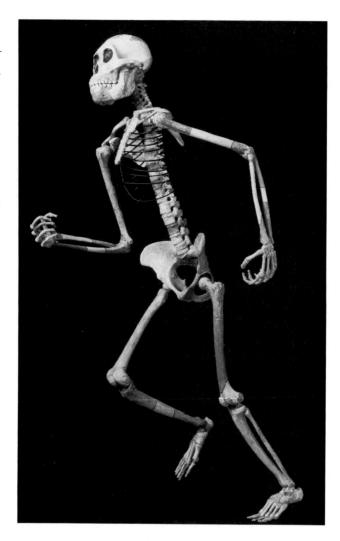

FIGURE 1.11

Restoration of the pelvis of
AL-288-1 (upper right) and its
comparison to those of a human
(lower right) and chimpanzee
(left). The restoration of this
specimen is based virtually
entirely on its preserved
morphology (i.e., it is not
reconstructed). The striking
overall similarity of the human
and australopithecine specimens is
evident.

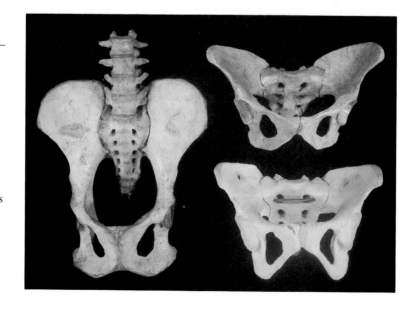

onstrates all the important anatomical features that distinguish human knees from those of quadrupedal hominoids, including a change in overall geometry that lowers stress in knee cartilage during complete extension of the joint (its position during upright walking). These same changes would also raise stress if the joint were used habitually in the flexed (**quadrupedal**) position.

The foot skeleton shows that the great toe, which is completely opposable in non-human hominoids and all other primates (allowing them to grasp branches and climb adeptly), had become permanently aligned with the other four toes (Latimer and Lovejoy, 1989, 1990a, 1990b; Latimer et al., 1987). Such an adaptation is important for bipedal walking, but it also represents the loss of the single most important hominoid anatomical adaptation for climbing.

Such specializations clearly imply that members of this species practiced bipedality as their primary form of locomotion and did so with such regularity that they rarely if ever used the arboreal canopy. This evidence is confirmed by changes that have taken place in their forelimb. When they adopted bipedality as a means of getting about, the entire burden of terrestrial locomotion fell exclusively to their hindlimbs — their forelimbs became emancipated from any significant role in terrestrial locomotion. If they had continued to use the arboreal canopy in a significant way, then we can reasonably surmise that not only would their forelimbs have remained adapted to arboreal locomotion, but they would have undergone further enhancement of that adaptation, because much of an early hominid's climbing ability would have been fatally compromised by changes in its hindlimbs. The shortening of the pelvis had made jumping and leaping more difficult; the specialized geometry of the knee would have increased the stress in this joint during climbing; and their foot had lost virtually all of its grasping ability. Were they to have attempted significant climbing bouts in the arboreal canopy, they would have been subjected to injurious and potentially fatal falls, had they not been anatomically compensated for the loss of these critical adaptations by an amplified capacity of the forelimb for climbing. And yet, early australopithecine forelimbs demonstrate a greater similarity to modern humans than to any other living hominoid. Their forelimbs were shorter, judged by comparison to their hindlimbs, than in any hominoid except modern humans (Figure 1.12), as were the digits of their hands. Clearly such arboreally important traits would have become amplified, not reduced, had the australopithecines been relying on climbing for either food or safety.

The australopithecine limb skeletons, therefore, provide two critically important clues. First, they show that early hominids rarely entered the arboreal canopy to satisfy any of their basic needs of food and safety (even humans occasionally climb, but as a species we do not rely on the arboreal canopy). Obviously, these ancestors had evolved effective mechanisms for terrestrial living, and their selection of bipedality, as a means of locomotion, provides us with a second important implication about them — that they must also have developed unique and specialized behavioral adaptations as well. We can deduce this from the basic nature of bipedality.

The Behavioral Significance of Bipedality

Even though bipedality restricted these early hominids to a terrestrial habitus, this odd form of locomotion was unlikely to have been a direct adaptation to terrestrial living as has so often been claimed. Both gorillas and chimpanzees are also primarily terrestrial, and when these large-bodied hominoids travel from one place to another they do so as terrestrial quadrupeds (**knuckle-walkers**). Their climbing abilities are used to exploit the arboreal canopy for food. They are arboreal feeders (Teleki, 1981).

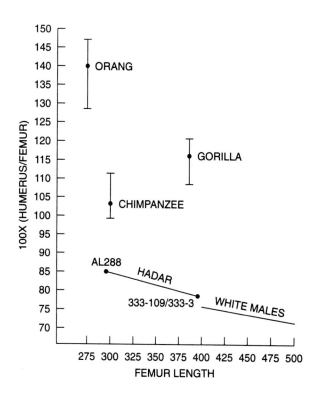

FIGURE 1.12

A plot of the humerus/femur ratio compared to femur length among hominoids. Only the average value of femur length is plotted for the ape samples (chimpanzee and gorilla: N = 20 each; orang: N = 11). For a larger sample of Caucasian males (N = 50), a regression line of the humerus/femur ratio and femur length is illustrated over the range of femur length observed in the sample. Note that the ratio is negatively correlated with femur length (as femur length increases, relative humerus length decreases). A data point for AL 288-1 ("Lucy") is plotted as well as reconstructed lengths for femur and humerus from a potential second single individual from Hadar (AL 333-109 [humerus] and AL 333-3 [femur]. Note that the humerus/femur ratio of Lucy is closer to those of small humans than to any other hominoid. Also note that there is a negatively allometric relationship that may partially account for the relative elongation of the humerus in *A. afarensis* relative to that in modern humans, because of the former's shorter femur.

The earliest hominids must have also been capable of a similar form of quadrupedal locomotion, and there would have been no inherent mechanical or energetic advantage to abandon it. In fact, there are many distinct disadvantages to bipedality. Contrary to popular opinion, it imparts no energetic advantage (walking bipedally requires the same energy expenditure as does walking quadrupedally [Taylor and Rowntree, 1973]); it greatly increases the probability of a debilitating injury (all loads are concentrated on two limbs rather than being more equally distributed on all four); and, as we have just discussed, its adoption greatly restricts a biped's capacity to exploit nonterrestrial environments. As bipeds we are slower, no more energetically efficient than quadrupeds, and at a much higher risk of musculoskeletal injury. Furthermore, the pelvis of AL-288-1 reveals yet another important selective force against bipedality — the pelvic modifications that it requires greatly restrict the size of the birth canal (Figure 1.13; Tague and Lovejoy, 1986).

A variety of **positional behaviors** have been proposed as selective forces for the emergence of bipedality, such as standing upright to see over vegetation or to pick foliage or fruit that would otherwise be out of reach. Yet there is *absolutely nothing preventing a primate quadruped from adopting such positions whenever they are*

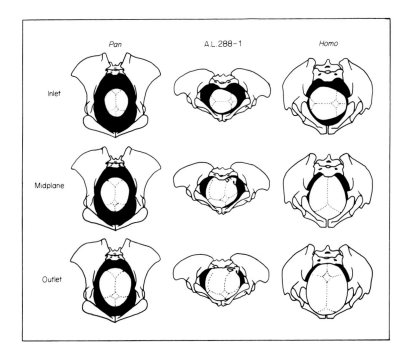

FIGURE 1.13

Comparison of birth mechanisms in a chimpanzee, AL-288-1, ("Lucy") and a modern human. For the former two species, a term fetal cranium from the common chimpanzee is illustrated, scaled to size. The dotted lines on the infant cranium represent its cranial vault sutures. In the chimpanzee, in which the birth canal is strikingly ample, no major reorientation of the fetal cranium is necessary during parturition. In the human specimen, rotation, flexion, and asynclitism (tilting) are required for passage of the cranium through the birth canal. These motions are also required for passage through the pelvic outlet (lower diagrams). In the australopithecine some of these motions would have also been necessary (especially rotation), but not all, making their birth process intermediate in difficulty between the relatively facile process in the chimpanzee and the much more difficult process in humans (Tague and Lovejoy, 1986).

needed. Chimpanzees and gorillas regularly do so for both feeding and social displays. There is no need, however, for them to adopt upright *locomotion* in order to habitually use upright posture. Some antelope species, such as the dik-dik, regularly feed for long periods in an awkward bipedal stance, but continue to use a quadrupedal (and extremely rapid) locomotor pattern.

The facts of bipedality are clear: it imparts no locomotor, energetic, or direct-feeding advantage to the species that adopts it; to the contrary it makes that same species more prone to injury, slower, and less able to exploit all of its environment (with the loss of climbing ability, much of a chimpanzee's food sources would be lost). It imparts no net energy advantage. In short, the adoption of bipedality by our early ancestors clearly was not a locomotor or postural adaptation.

Together, these facts imply that bipedality must have been some form of behavioral adaptation, and that it was adopted when a loss of access to the arboreal canopy was not a critical handicap. In other words, it must have developed in a hominoid whose resource base was largely, if not exclusively, already terrestrial. This is not particularly surprising in light of what we now know about other hominoids. Recent data suggest that large-bodied, terrestrial hominoids were also part of the Asian radiation (the only survivor of which is the orangutan; [Pilbeam et al., 1990]). Some ancient Asian apes, such as *Gigantopithecus,* were so large that they were almost certainly restricted to terrestrial resources.

Occupation of the forest floor would have required a shift in primary diet. Many of the various fruits, shoots, and young leaves that chimpanzees obtain using their acrobatic arboreal skills would have been unavailable to terrestrial hominids. They could have responded by increasing body size and relying on high quantities of low-quality nutrients—a strategy adopted by the largely terrestrial gorilla. However, this would not explain the dramatic shift toward omnivory in the early hominid dentition and it is inconsistent with their estimated body size. Instead, given the combination of their dental and locomotor specializations, early hominids would appear to have been forest floor generalists, relying on eclectic and varied resources as do other terrestrial mammalian omnivores.

In summary, the evidence provided by their bones and teeth suggests that the earliest hominids probably exploited a variety of relatively high-energy food sources, that their adaptation to such a niche was already complete during the time in which they adopted bipedality (there must have been little or no selection against the loss of climbing ability), and finally, that their actual adoption of upright walking was in response to some behavioral advantage, and not to any locomotor or postural need.

Some Further Projections about Early Hominid Adaptations

Several additional conclusions naturally follow if we apply some basic rules of ecological adaptation in mammals and primates. First, early hominids must have used a search-intensive feeding strategy more like that of some Old World monkey species (such as some macaques) than the handling-intensive strategy used by other primates that rely on low-quality foods (such as the gorilla or some baboons). High-quality food items tend to be more thinly and cryptically distributed in an environment than bulk, leafy plant foliage (Pianka, 1974). As we have seen, early hominids lacked both the large body size and masticatory shear that would be necessary to exploit the latter regimen, and their terrestriality would have eliminated tree-borne fruit as a staple food supply. They are clearly much more likely to have relied on a diet of forest-floor, forest-edge, lake-margin, and riparian (stream-edge) sources of dispersed but highly nutritious items such as insects, annelids, fledglings, eggs, fish, amphibians, small reptiles, and other invertebrates, as well as some young shoots, tubers, and lower-canopy fruits (Mann, 1981). Such search-intensive omnivory as well as the restriction to ground dwelling would both have required a considerably larger home range (Jolly, 1985; Clutton-Brock and Harvey, 1978).

The picture of early hominids that emerges from a systematic review of their anatomical features is perplexing in light of their success as a group relative to other hominoids. Despite having been restricted to virtually exclusive terrestriality, a search-intensive food strategy, and a slow, awkward locomotor pattern, early hominids still managed to penetrate novel habitats never successfully occupied by other hominoids and to favorably compete with a variety of Old World monkeys. By the **Pliocene**, hominids had occupied and become plentiful in a variety of African habitats, including the savannas of Laetoli, Tanzania, the lake margins of Northern Ethiopia and Kenya, and the travertine caves of the high veldts of South Africa. Hominids most probably occupied the forest floors of West Africa as well, although there is little chance of **paleontological** confirmation of this because of the shortage of fossil-bearing sediments of the right age exposed in West Africa today. By about 1 million years ago, the descendants of these early hominids had probably invaded many terrestrial habitats in Eurasia and Africa.

Why were our slow, awkward ancestors so successful while other hominoids, which occupied food-rich forest canopies for millions of years, were not? And why was the success of these early hominids coupled with the appearance of a relatively bizarre form of locomotion? These questions are fundamental to any attempt to construct a scenario of early hominid evolution. There are, fortunately, a plethora of clues to answer these questions.

Two Traditional Explanations for Bipedality

Two traditional explanations for bipedality have been negated by developments in the fossil record during the last decade. The first and most traditional answer, that upright walking was a response to the emergence of stone tool making, has not been borne out by the fossil record. First, the oldest stone tools, which are exceptionally primitive, have been found to be only about 2.5 million years in age. Fully developed bipedality is at least 1.5 million years older. It is sometimes suggested that this gap was filled by a reliance on tools made of perishable material, and that tool use in early hominids must predate the earliest stone tools by a substantial margin. Indeed, perishable material almost certainly preceded the first use of stone as a raw medium for tool making. Crude tools may have played a significant role in the search-intensive feeding strategy of early hominids, and may have been used to provide access to some foods that would have otherwise gone unexploited. Yet it is unlikely that tools made of simple, readily available materials were so pivotal to the earliest hominids that they would have developed a specialized form of locomotion merely to carry them from one point to another when they could be easily fashioned anywhere on the forest floor.

The first use of stone was a critical breakthrough because it would have been the first sufficiently durable material with which a variety of important tasks could be accomplished. Less suitable media, however, are very unlikely to have played the pivotal role in early hominid evolution that **lithic** technology would eventually assume (see Potts's discussion in Chapter 3).

A second character once believed to be associated with the emergence of bipedality is the substantial enlargement of the hominid brain. It has often been presumed that greater intelligence may have been somehow directly associated with the adoption of upright walking. Again, however, the fossil record is quite explicit in decoupling the emergence of bipedality from any substantial increase in brain size. All indications currently suggest that the brain size of *A. afarensis* was (relative to body size) quite close to that of both modern chimpanzees and gorillas. So why did the ancestors of *A. afarensis* adopt upright walking, and why was it coupled with their exceptional success in competing with other primates?

■

THE REPRODUCTIVE ADAPTATIONS OF HOMINIDS

The most direct way of addressing this question is to turn to the single most critical factor in the success or failure of any species, its reproductive "strategy." This is a critically important aspect of all hominoids because they are among the most slowly reproducing of all mammals. Their exceptionally slow reproductive rate is largely due to their slowly maturing infants that require intense parental care to acquire the social skills and environmental knowledge needed to compete successfully with conspecifics.

Such extreme parental investment has profound **demographic** consequences. Birth spaces of 4 to 5 years require each female chimpanzee to survive to an average age of about 18 to 20 years merely to maintain the population into which she is born. Animals with such severe demographic specialization are expectedly rare because such

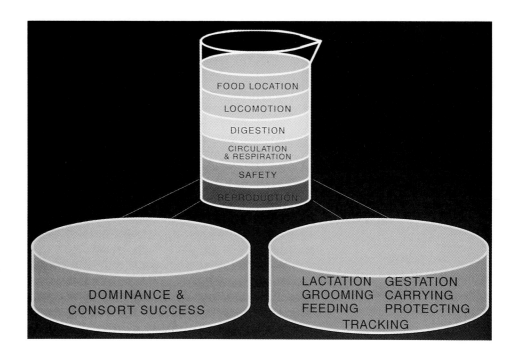

FIGURE 1.14

A hypothetical beaker of energy is displayed with layers devoted to the required physiological and behavioral activities in the life of a successful mammal. Each of these activity layers can be attributed to one of three basic modes, together referred to as the fundamental selective triad (see text): food, safety, and reproduction (obviously, some layers must be apportioned among these three fundamental modes). Only the layer attributed to reproduction differs substantially in male and female primates, and as illustrated, the individual activities differ dramatically in the two sexes. Note that male reproductive energy is typically spent in competition with other males, a phenomenon that results in no substantial contribution to offspring survivorship. In contrast, virtually all of female reproductive energy is devoted to improving the probability of survivorship of her infants. See also Figure 1.15.

low reproductive rates make it exceptionally difficult to respond successfully to substantial environmental change (such as during the Pliocene) or to the emergence of new competitive groups (such as Old World monkeys).

There is, of course, a striking exception—our hominid ancestors. The Pliocene environment in which they lived was rapidly changing, and they, too, had to compete with many terrestrially adapted Old World monkeys. Their success is likely to have involved a significant increase in their reproductive rate, but this would have in turn required some substantial shift in reproductive adaptations. Moreover, such a shift must have been coupled with those anatomical characters that we have already described: bipedality, forest-floor omnivory, large home ranges with a search-intensive feeding strategy, and canine monomorphism. Can such a coupling be made and does it provide a solution to the mystery of early hominid ecological success?

The Fundamental Selective Triad

Two avenues of improvement in reproductive rate would have been open to late Miocene hominoids: either a reduction in the mortality rate of their young, or a substantial reduction in the birth space between offspring (or both). A shorter period of infant dependency was not possible, as it would have attenuated the learning period

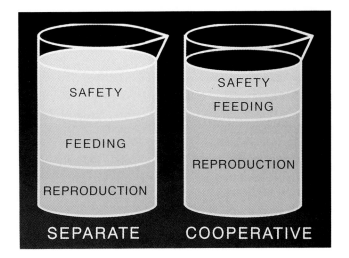

FIGURE 1.15

Fundamental selective triad beakers of females operating in primate societies that lack male-female cooperative activity in improving offspring survival (*separate*), and in those that are *cooperative*. The figure illustrates that such cooperation does not necessarily involve direct paternal care of offspring (such as is observed in callitrichids). Rather, reduction of one or both nonreproductive elements in the female's fundamental selective triad causes the energy apportioned to reproduction to be increased, thus favoring both the female and her cooperating male. Provisioning behavior in mammals and birds causes a reduction in the amount of daily energy required to achieve both food and safety. Surplus energy made available to the female as a consequence of provisioning is thus transferred to reproduction, thereby increasing the fitness of both partners.

of the infant and made it less competitive and capable of survival. Yet the effective birth rates of hominids must have somehow been increased. This has to mean that females could reproduce a second or third offspring before the first was fully independent. Having multiple dependent offspring is very unusual among primates. It requires enormous investment on the part of the mother, and yet this evolutionary transition from having one to several dependent offspring certainly occurred sometime in the evolution of our species. How could a female hominoid afford to take on this taxing reproductive regimen?

Figure 1.14 may hold a potential solution to this dilemma. A beaker is shown which, for the purposes of discussion, can be imagined to contain a full day's energy supply. The various layers in the beaker represent different activities on which any organism must spend that energy during its daily life cycle.

Note that all of the various physiological, neurological, and social activities shown in the first beaker can be concentrated into three essential areas of energy expenditure. These we call a *fundamental selective triad*. Quite simply, if any organism can meet all nutrient requirements, avoid mortality (by disease, accident, or predators), and reproduce at a high rate, then it is almost certain to thrive and compete successfully not only with its conspecifics but with all other organisms as well. Such an animal is an evolutionary ideal. Obviously no such "perfect" mammal or primate exists, but it is instructive to note the three most basic areas of energy expenditure: nutrition, safety, and reproduction. Figure 1.15 demonstrates that should the requirements in one of these categories be more easily met, expenditures in the others can be correspondingly increased. Of course, safety and nutrition have obvious natural maxima. There are limits to the food and safety any organism requires. There is no such ceiling, however, on the value of reproduction. The more any mammal can successfully reproduce, the greater proportion he or she contributes to future generations—this is the fundamental basis of natural selection.

The Fundamental Selective Triad in Primates

As a rule, male primates contribute little or nothing to the female's fundamental selective triad. Their reproductive role is often little more than supplying genetic variation via sperm, and although they indirectly may contribute to food and safety by repelling predators or defending a territorial boundary from conspecifics, they often also decrease offspring safety by their aggressive behavior with one another (or even by practicing infanticide), and reduce food supplies by consuming their group's resources (Hrdy, 1977).

In the fundamental selective triad, a transfer of any energy that can be spared from feeding and safety to reproduction will be highly favored by natural selection. A prominent avenue by which such shifts can occur is available to most primates. This is to use previously untapped male energy to contribute directly or indirectly to the female's fundamental selective triad. In a minority of primates and mammals (but in most bird species), such a shift has taken place. In various species of **callitrichids**, for example, males perform many of the tasks of infant care. This allows the females of these comparatively tiny primates consistently to twin, which doubles their reproductive rate. In many other primate species such as owl monkeys and gibbons, parental care is more equally shared—but in each species an important increase has been effected in the resource base of reproduction. In most bird species both parents provide direct care of the offspring. Why haven't more mammals and primates made this highly advantageous switch?

The answer, as always, lies in the nature of natural selection. Selection favors the most reproductively successful individuals, not necessarily those structures and behaviors that have the greatest benefit for the species as a whole. As individual infants come to require increased levels of care and become rarer (that is, the birthspace becomes greater), the total number of offspring born to an individual female is reduced, and the survivorship of each becomes proportionately more critical to her overall success. When reproductive rate becomes sufficiently slowed—the success or failure of a single offspring may ultimately differentiate between a highly successful and a relatively unsuccessful female.

The same demographic constraints and parameters operate on males as well. Aggressive competition with other males is an effective reproductive adaptation only so long as each female's reproductive rate remains sufficiently high to provide reproductive opportunities to the male. If her potential rate falls sufficiently low, so will the number of opportunities to impregnate her. The rarity of females who are not pregnant or nursing is directly proportional to the length of the time elapsed between births. In hominoids such females are very rare. Because they are so uncommon, selection favors those males who can contribute more energy (either directly or indirectly) to their own offspring's survivorship.

Provisioning Behavior in Miocene Hominids

With this as background, it is time to reconsider the forebears of the australopithecines. Their success testifies to their having undergone a fundamental shift in reproductive adaptation not observed in other (living) hominoids. The clues to the nature of this shift have already been presented. We are in search of some change in reproductive adaptations that could allow male behavior to improve the allocation of energy within a female's fundamental selective triad. Such a shift would have to occur within the context of a terrestrial, search-intensive, omnivorous feeding strategy with expansive home ranges and minimum territoriality, and is in some way probably related to the adoption of bipedality.

Stated within this context, one particular solution seems to stand out. Bipedality does provide one particular advantage over quadrupedal locomotion: it frees the hands for carrying. Provisioning—the location, collection, and furnishing of food— is a common reproductive adaptation in many mammals and most bird species. Bipedality, even though of no particular energetic, positional, or mechanical value, would allow the carriage of food over significant distances (Hewes, 1961).

Suppose that female hominids began to mate selectively with males who occasionally provisioned them. She would enjoy several important advantages. First, there would be a proportional reduction in the food component of her fundamental selective triad. Moreover, in a search-intensive feeding regimen, her path length would be substantially reduced as search time was shifted to a male mate, even if he were only temporary. Such reduction would improve the survivorship of her young, since it would allow her to occupy a smaller and more familiar core area.

The potential significance of a provisioning strategy is, of course, dependent on the extent to which such behavior could have played a significant role in fundamentally shifting the makeup of the fundamental selective triad—that is, how potentially valuable would male provisioning have been to an early hominid female? Ecological studies of living primates can provide this kind of information.

An instructive example is one conducted on a species of Old World monkey, the toque macaque, *Macaca nemestrina* (Hladik, 1975). Hladik first carefully plotted the limits of the home range of a target troop. He then identified and mapped the position of each tree in the home range and calculated the potential caloric value of its product as consumed by the macaques. In doing so he found that available fruit exceeded that required to meet annual caloric needs by a factor of one hundred. Why, then, do these primates defend so large a home range?

The answer is related to the other dietary needs that each animal must fill. In addition to calories, each member of the troop (and especially the females and young) must also obtain vitamins, fats, fatty acids, and proteins (amino acids) necessary for growth and reproduction. Fruit supplies few of these nutrients. They must be obtained by supplementing with foods that contain significant amounts. Such foods are largely either nuts or some form of animal protein. Hladik found that most macaque search time was spent seeking these rarer, and more dispersed, sources of fats and proteins.

Were female hominoid requirements for such supplementation sufficiently great to selectively favor such a change in feeding adaptations; that is, to cause the emergence of actual male provisioning of females? An indication of its potential value is provided by data collected by Jane Goodall at Gombe (Goodall, 1986). Figure 1.16 shows the average total proportion of feeding time spent by female chimpanzees in search of insects (which are exceptionally high in dietary fat and protein). Such data are quite striking. More than 50% of annual feeding time invested by females is spent specifically searching for these dietary supplements. Yet each individual insect contributes only slightly to the nutrient supply of each chimpanzee. It is their collective impact that is important.

Consider the potential role of provisioning behavior in a hominoid that was already a forest-floor dweller and in which a search-intensive strategy had already evolved (that is, both sexes met their nutritional needs by such behavior). The location of critical food items would have been already a regular part of the male's behavioral repertoire, and the proposed shift in overall reproductive adaptations would have required only that males offer and provide occasional food items to potential

FIGURE 1.16

Total feeding time devoted to searching for insect prey by female chimpanzees at Gombe (average of monthly data for a period of 2 years). Data from Goodall, 1986.

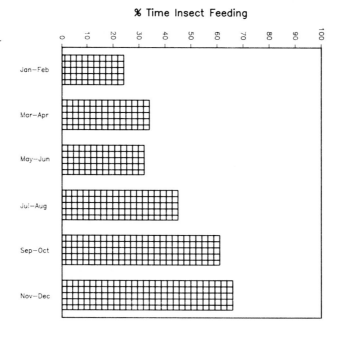

% Time Insect Feeding

mates, a practice commonly seen today among bonobos (Kano, 1982). Fixation of such behavior would, in turn, have required only that it have more impact, as a reproductive strategy, on the breeding success of its participants than alternative strategies (such as male competition for females in estrus). What evidence might we seek that could indicate such a shift in reproductive adaptation played a central role in early hominid evolution?

The Epigamic Adornment of Hominids

Any male who practiced such provisioning behavior would, of necessity, have done so only for a female who bore *his* offspring. In the early stages of the emergence of provisioning, males therefore would have probably used some form of restrictive mating behavior such as **consortships** seen today in common chimpanzees. When female chimpanzees enter **estrus** they regularly copulate with several males. This greatly reduces the chance that any particular male will father an offspring. Male chimpanzees, as a consequence, have developed very large testicles, which produce copious amounts of sperm. An even more successful reproductive adaptation among these same male chimpanzees is to urge and guide a female in the early stages of estrus away from the group, so that when she does ovulate (at mid-estrus), he will be the only male with which she copulates. This is apparently an effective trait that enhances reproductive success among male chimpanzees. Tutin (1979, 1980), for example, has calculated that although only 2% of chimpanzee copulations occur within the context of such consortships, more than half of successful copulations occur during them.

Such a temporary pairing with a provisioning male in early hominids would probably only have marginally increased the reproductive success of the participating female. However, given the almost certain rarity with which males were confronted with fertile females, and given the excessive value of each offspring, any extension of provisioning behavior beyond the immediate context of a consortship could have proportionately increased the reproductive success of both participants: the greater the probability of a successful pregnancy and birth, the greater the reproductive fitness of both members of the temporary pair. Selection would therefore favor females who chose males on the basis of the duration of provisioning behavior. It would have forced males, on the other hand, to choose females who were unlikely to bear the offspring of other males. We can therefore make two further predictions about early hominids.

First, we would expect both sexes to be *epigamically decorated*. An **epigamic** feature is a visible anatomical structure associated with sexual identification, but which is not itself directly involved in the reproductive process. Such characteristics regularly appear in members of the sex that is the subject of active choice by the other, but are normally absent if such choice is also absent. The most typical examples occur among bird species, in which males compete with one another for selection by females. In such species, those characteristics on which females rely for mate selection become dramatically elaborated, such as color, tail length, pelage patterns, and so on (Mayr, 1963).

Epigamic features usually have nothing directly to do with physical reproductive success (although they are often anatomically associated with the reproductive process). Mate choice is a genetically determined character of the female, which her female offspring inherit. The male anatomical attributes that are the object of her choice are, of course, also genetically determined, and her male offspring inherit them. Elaboration of such anatomical attributes therefore occurs in the sex that is the object of the other's choice. The most successful pairings are those in which the anatomical features of choice are also associated with those that physically improve reproductive rate, even though the latter are rarely the actual basis of such choice. For example, in early hominids there would have been virtually no opportunity for females to choose males on the actual basis of their reliability and skill in provisioning, but invariably those females who did choose such males would be the most reproductively successful. Whatever anatomical features on which they physically based their actual choice would then become the most selectively favored and would be accentuated.

Most male primates are only moderately decorated with epigamic features. Almost no female primates are. Therefore, the elaborate epigamic adornment of both sexes in our own species tells us a great deal about our hominid ancestry. Human males exhibit scalp and facial manes, a large, pendulous penis (relative to that of other hominoids), specialized organs of scent (composed of both **sebaceous glands** and hair) in the axillary and pubic regions, a distinct voice, and a body size approximately 20% larger than that of females. Human females also exhibit scalp manes, a distinct voice, and the specialized scent organs of the **axilla** and **pubis**. In addition, they display a unique (among primates) distribution on the body of fat, and nonlactational breast enlargement (a singular feature among primates in which breast size normally cycles demonstrably with **lactation**).

The most unusual characteristic of human females, however, is their failure to exhibit any external anatomical or behavioral sign that **ovulation** is occurring. As most female primates enter estrus, a series of behavioral and visible anatomical changes denote their special status of being imminently impregnable. Because such signals attract males they greatly increase the probability of active sperm being present when ovulation occurs, thus assuring pregnancy. In some species in which little or no competition among males occurs, such external signs of ovulation are reduced and replaced with active solicitation of copulation by females, a phenomenon termed *proceptivity*. In virtually every monkey and ape species known, one of these two means of assuring successful copulation during ovulation is observed. The one exception appears to be our own species. In humans, the occurrence of ovulation is hidden from both sexes, and females show neither special anatomical nor behavioral changes during ovulation. This is a striking attribute because it directly implies that copulation among humans is sufficiently regular and frequent to assure fertilization. Otherwise, signals of active ovulation would be selectively favored. Are these highly unusual anatomical and behavioral features of humans related to the reproductive behavior that developed in our australopithecine ancestors? There is a good chance that they are.

If provisioning behaviors were to have been fixed by selection in a hominoid species with very long birth spaces, what epigamic and behavioral choice variables would we predict to have been established? Because, as we have noted above, such an adaptation involves both sexes contributing (directly by females, indirectly by males) to the survivorship of offspring, we would expect both sexes to exhibit active mate selection—their reproductive success would vary according to the quality of their mate. Let us first consider such mate selection by females. What anatomical features would males exhibit?

The Special Epigamic Characters of Early Hominid Males

Male primates frequently compete for relative status within defined dominance hierarchies. Their position is dependent on their physical ability to dominate other males either by threat, combat, or a combination of the two. As a result, males in such species demonstrate relatively large body size (compared to that of their female counterparts) and enlarged canine teeth. Among chimpanzees, however, whose basic social structure is probably more representative of the early hominoids from which hominids emerged, there is considerably less competition among the males of a single group because chimpanzee males tend to be related. This is a direct consequence of their being a *female transfer* species—at adulthood females leave their natal group and take up residence elsewhere in another, while males usually remain in the group in which they were born. This leads to a moderate degree of relatedness among such males.

Chimpanzee males exhibit a relative dominance hierarchy, and this is sometimes reflected in their reproductive behavior. Very dominant males tend to copulate more frequently with females who are in estrus than do subordinate males. The latter display a greater tendency to attempt restrictive mating relationships such as the consortships described above.

Early hominid females who selected males with reduced canines would therefore be choosing males who were also more likely to exhibit restrictive mating and any associated behaviors such as provisioning. Males at the apex of a dominance hierarchy (such as those with the largest and most threatening canines) would be most likely to continue to achieve reproductive success by dominating females who had entered estrus. Such males would, however, be the least likely to exhibit significant provisioning behavior. Conversely, males who had the lowest probability of reproductive success with fully estrous females would be most likely to attempt restrictive mating and to entice females with alternative behavioral patterns. *In short, the probability of significant moderate-term provisioning behavior would be inversely related to canine size.* This would explain the striking reduction of canine crown size in the earliest australopithecines.

The Special Epigamic Characters of Early Hominid Females

What physical attributes would have formed the basis for male choice? The most critical factor for mate selection by a male would be any characters that would assure

FIGURE 1.17

Frequency of copulation in chimpanzee females as a function of physiological status. Note pronounced reduction of copulation during lactation. After Lemon and Allen, 1978.

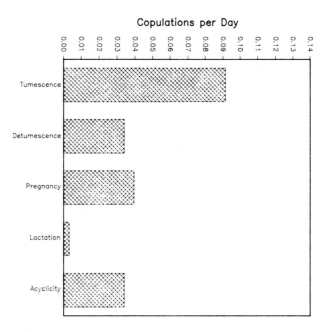

him that any offspring born to his mate would be genetically his. Just as the most successful females would have been those who chose males without massive canines (those not associated with an aggressively competitive male strategy), the most successful males would have been those who chose females who were least attractive to aggressive males — those with little or no advertisement of actual or impending ovulation.

In addition to explaining the absence of external signs of ovulation in their descendants, this also provides an explanation of another unique human character. Human females display the remarkable feature of nonlactational breast enlargement. Among other primates, breast size cycles with lactation. Because of lactational **amenorrhea** (the feedback mechanism whereby breast stimulation by a nursing infant prevents ovulation), enlarged breasts serve as prominent physical symbols that a female is not ovulating and is thereby not available for impregnation. This is strikingly reflected in the copulation frequencies seen in chimpanzees (Figure 1.17; Lemon and Allen, 1978). Male hominids who were mated to females who retained enlarged breasts even though they had reentered ovulatory cycling would have been most strongly protected from potential cuckoldry from other males. The same condition would hold, of course, as an explanation for the loss of any external advertisement of ovulation. Natural selection would favor those males who provisioned and regularly copulated with females whose reproductive status remained cryptic — hidden from other males. Regular copulation would be necessary to assure fertilization once the female reentered her ovulation cycle, and it would also serve, as it does in other species, as a social signal of a particular pair-bond between the participants — it would function to maintain the bond that had become pivotal to the success and survival of the offspring of the mated pair.

In addition to the selection of females who displayed reproductive crypsis, males would also be expected to select females of the highest **fecundity** — those most likely to be reproductively successful. Inasmuch as fat stores are one of those physiological attributes most associated with reproductive success (Frisch, 1975, 1988), a prominent display of such stores, as is evidenced in the special female human **somatotype**, would also be an obvious correlate of provisioning adaptations.

In sum, males whose reproductive adaptations involved the active provisioning of a female would be expected to bond sexually with such a female on a moderate to long-term basis. With birth spaces in excess of 3 to 4 years, provisioning behavior would have only become dominant if it were practiced over a significant period of time—in particular if the male's behavior were to have shortened the birth space by accelerating the weaning of the offspring such that the mother would re-enter her ovulatory cycle. This, then, would have been the pivotal adaptive feature of a provisioning strategy—not only would offspring enjoy a richer (more nourishing) and more stable early developmental period (with less exposure to mortality risk), their mothers would have re-entered ovulatory cycling at an earlier point in their maturation—thus shortening the birth space. Those males who had pair-bonded and had continued to copulate regularly with a provisioned female (whose reproductive status remained cryptic) would have exhibited a high probability of fathering a second offspring at the earliest available window of female fertility. Not only would infant survivorship have been enhanced, but the length of each birth space would also have been significantly reduced. This shift in behavior would have directly favored females who mated with competent, reliable males, and males who mated with fertile, reproductively cryptic females. A review of all the epigamic features of our species reveals that precisely those anatomical signals are specifically expected within such a reproductive adaptation.

THE POST-AUSTRALOPITHECINE PERIOD OF HUMAN EVOLUTION

The fossil record is clear in recording the major events of human evolution subsequent to those of the earliest australopithecines. At least two distinct australopithecine lineages emerged, usually allocated to two distinct groups known as the *robust* and *gracile* forms, based primarily on the degree of postcanine dental expansion and associated changes in their facial and masticatory anatomy. The robust australopithecines probably were the result of exploitation of the more resistant, protected food items found in a drier woodland savanna and open grassland niche; the more *gracile* forms appear to have remained more dietarily generalized and to have continued to inhabit a broader variety of environments.

The origin of the genus *Homo* is still unclear, with a plethora of taxonomic scenarios currently in debate. Although some researchers adopt the view that the genus *Homo* emerged from the gracile australopithecine **clade**, others contend that it represents an independent development directly from *A. afarensis* (that it evolved in parallel with the later australopithecine lineages). This debate will likely not be resolved until the fossil record is substantially improved (see Walker's discussion in Chapter 2).

There are many other aspects of the emergence of our direct ancestors of the **Plio-Pleistocene**, however, that are in considerably less doubt. It is a virtual certainty, for example, that the brain of post-australopithecine hominids underwent a dramatic and rapid expansion, that members of early *Homo* radiated widely throughout the Old World, and that lithic technology continued to emerge, but at an exceedingly slow pace (Potts, Chapter 3).

The remarkable success of early *Homo* must not be forgotten as a major item of evidence bearing on scenarios of the earliest phases of human evolution. The **cerebral cortex** is useful primarily as a device for the storage and retrieval of information. Its later expansion during the Pleistocene is direct evidence of a highly advanced reproductive adaptation that would allow further protraction of the time period of subadult dependency; that is, substantial mechanisms assuring high levels of survivorship must have been in place by the beginning of the evolutionary radiation of

Homo. The roots of these social adaptations must surely have stretched back in time into the preceding australopithecine phase of human evolution, as it was in this phase that the early hominid adaptive radiation was initiated.

■

SUMMARY: AN ORIGINS SCENARIO AS A FORMAL HYPOTHESIS

This chapter begins with the analogy of a criminal trial to introduce the concept of a formal "scenario" and how its probability and accuracy could be judged. Although space prevents mention of all the relevant evidence of early human evolution as well as a complete discussion of the facts presented in this chapter, this evidence leads to the formulation of a scenario (Lovejoy, 1981), a synopsis of which follows:

The earliest hominids became a distinct evolutionary lineage separate from that of other hominoids during the latter phases of the Miocene hominoid radiation (approximately 7 to 9 million years ago). At the time of their emergence they occupied a largely non-arboreal, forest floor niche, and had an omnivorous, search-intensive diet, which included a significant intake of calorie- and protein-rich meat and animal fat. Home ranges, core areas, and daily travel distances were all greater than those of more arboreal hominoids. Social structure was similar to that of chimpanzees and bonobos, with female transfer and male bonding within groups of approximately equal sex ratios. Typical sexual behavior included promiscuity, moderate male dominance hierarchy, and restrictive matings commonly seen in chimpanzees and bonobos today. With increased protraction of the average birthspace (encouraged by conspecific competition for maximum behavioral plasticity), restricted mating pairs in which males continued to provide food items of high fat and/or protein content into the post-ovulatory phase began to yield greater net reproductive output for the engaged males than did more typical male reproductive behavior. The reproductive success of such pairs was enhanced by reduced infant mortality (smaller daily path length and core area; more reliable protein and fat supply) and intensified female parenting (shift of energy within the fundamental selective triad as a consequence of male provisioning). The males of such pairs were most successful if competently bipedal and capable of proficient provisioning. Such males were also more successful if paired with females who expressed ovulatory crypsis (reduced external signs of ovulation and non-cyclic breast enlargement) and demonstrable fat stores (as exhibited by distinctive somatotype). Females were most successful if paired with males that exhibited reduced canine crowns, because such males were least likely to attempt sexual success by a strategy of aggression within the dominance hierarchy. Active choice by both sexes led to elaborate epigamic decoration of both by behavioral linkage disequilibrium. Enhanced survivorship and reduced subadult mortality allowed invasion of novel habitats and successful competition with terrestrial Old World monkey species. The further reduction of intra-group male-male aggression made possible by pair-bonding within a single group led to enhanced group cohesion. This specialized set of attributes led to the radiation of australopithecine species into new habitats (in which at least one speciation event occurred) and the emergence of the genus *Homo*.

The final question for the reader of this essay is the degree to which the hypothetical scenario just presented *logically accounts for each item of available evidence by using a minimum of special arguments and a maximum of explanatory economy.* The most authentic scenario of human evolution should account for what appear to be exclusive or specialized human traits by the same selective agents implicated in the evolution of other mammalian and bird species. The final verdict as to the success of that presented here is the exclusive prerogative of the reader.

SUGGESTED READINGS

Conroy, G. C. 1990. *Primate Evolution* (New York: Norton).

Fleagle, J. G. 1988. *Primate Adaptation and Evolution* (New York: Academic Press).

Goodall, J. 1986. *The Chimpanzees of Gombe: Patterns of Behavior* (Cambridge, Mass.: Belknap Press of Harvard Univ.).

Johanson, D. C., and Edey, M. 1981. *Lucy: The Beginnings of Humankind* (New York: Simon and Schuster).

Lovejoy, C. O. 1988. Evolution of human walking. *Scientific American 259*: 118–125.

REFERENCES

Andrews, P. 1981. Species diversity and diet in monkeys and apes during the Miocene. In: C. Stringer (Ed.), *Aspects of Human Evolution* (London: Taylor and Francis), pp. 25–61.

Bernor, R. L. 1983. Geochronology and zoogeographic relationships of Miocene Hominoidea. In: R. L. Ciochon and R. C. Corruccini (Eds.), *New Interpretations of Ape and Human Ancestry* (New York: Plenum Press), pp. 21–66.

Clutton-Brock, T. H., and Harvey, P. H. 1978. Mammals, resources, and reproductive strategies. *Nature 273*: 191–194.

Frisch, R. E. 1975. Demographic implications of the biological determinants of female fecundity. *Social Biology 22*: 17–22.

Frisch, R. E. 1988. Fatness and fertility. *Scientific American 258* (3): 88–95.

Goodall, J. 1986. *The Chimpanzees of Gombe: Patterns of Behavior* (Cambridge, Mass.: Belknap Press of Harvard Univ.).

Goodman, M., Tagle, D. A., Finch, D. H. A., Bailey, W., Czelusniak, J., Koop, B. F., Benson, P., and Slightom, J. L. 1990. Primate evolution at the DNA level and classification of hominoids. *Journal of Molecular Evolution 30*: 260–266.

Hatley, T., and Kappelman, J. 1980. Bears, pigs, and Plio-Pleistocene hominids: A case for the exploitation of below ground food sources. *Human Ecology 8*: 371–387.

Hewes, G. W. 1961. Food transport and the origins of hominid bipedalism. *American Anthropologist 63*: 687–710.

Hill, A., and Ward, S. C. 1987. The origin of the Hominidae: The record of African large hominoid evolution between 14 million years and four million years. *Yearbook of Physical Anthropology 31*: 49–83.

Hladik, C. M. 1975. Ecology, diet, and social patterning in Old and New World primates. In: R. Tuttle (Ed.), *Socioecology and Psychology of Primates* (The Hague: Mouton), pp. 3–35.

Hrdy, S. 1977. *The Langurs of Abu* (Cambridge, Mass.: Harvard University Press).

Johanson, D. C., and White, T. D. 1979. A systematic assessment of early African hominids. *Science 203*: 321–330.

Johanson, D. C., Taieb, M., and Coppens, Y. 1982. Pliocene hominids from the Hadar Formation, Ethiopia, 1973–1977: Stratigraphic, chronologic, and paleoenvironmental contexts, with notes on hominid morphology and systematics. *American Journal of Physical Anthropology 57*: 373–402.

Jolly, A. 1985. *The Evolution of Primate Behavior,* 2d ed. (New York: Macmillan).

Kano, T. 1982. The social group of pygmy chimpanzees (*Pan paniscus*) of Wamba. *Primates 23*: 171–188.

Kelley, J., and Pilbeam, D. 1986. The dryopithecines: Taxonomy, comparative anatomy, and phylogeny of Miocene large hominoids. In: D. R. Swindler and J. Erwin (Eds.), *Comparative Primate Biology, Vol. I, Systematics, Evolution, and Anatomy* (New York: Alan R. Liss), pp. 361–411.

Latimer, B., and Lovejoy, C. O. 1989. The calcaneus of *Australopithecus afarensis* and its implications for the evolution of bipedality. *American Journal of Physical Anthropology 78*: 369–386.

Latimer, B., and Lovejoy, C. O. 1990a. The hallucal tarsometatarsal joint in *Australopithecus afarensis*. *American Journal of Physical Anthropology 82*: 125–133.

Latimer, B., and Lovejoy, C. O. 1990b. Metatarsophalangeal joints of *Australopithecus afarensis*. *American Journal of Physical Anthropology 83*: 13–23.

Latimer, B., Ohman, J. C., and Lovejoy, C. O. 1987. Talocrural joint in African hominoids: Implications for *Australopithecus afarensis*. *American Journal of Physical Anthropology 74*: 155–175.

Lemon, W. B., and Allen, L. A. 1978. Continual sexual receptivity in the female chimpanzee (*Pan troglodytes*). *Folia Primatologica 30*: 80–88.

Lovejoy, C. O. 1981. The origin of man. *Science 211*: 341–350.

Lovejoy, C. O. 1988. Evolution of human walking. *Scientific American 259*: 118–125.

Mann, A. E. 1981. Diet and human evolution. In: R. S. O. Harding and G. Teleki, *Omnivorous Primates Gathering and Hunting in Human Evolution* (New York: Columbia Univ. Press), pp. 10–36.

Mayr, E. 1963. *Animal Species and Evolution* (Cambridge, Mass.: Belknap Press of Harvard Univ.).

Pianka, E. R. 1974. *Evolutionary Ecology* (New York: Harper and Row).

Picq, P. 1991. The diet of *Australopithecus afarensis*: An attempted reconstruction, [translated by I. Tattersall]. In: E. Delson, I. Tattersall, and J. Van Couvering (Eds.), *Paleoanthropology Annuals, Vol. I, 1990* (New York: Garland), pp. 99–102.

Pilbeam, D., Rose, M. D., Barry, J. C., and Shah, S. M. I. 1990. New *Sivapithecus* humeri from Pakistan and the relationship of *Sivapithecus* and *Pongo*. *Nature 348*: 237–239.

Sibley, C. G., and Ahlquist, J. 1987. DNA hybridization evidence of hominoid phylogeny: Results from an expanded data set. *Journal of Molecular Evolution 26*: 99–124.

Steininger, F. F., Rabeder, G., and Rogl, F. 1985. Land mammal distribution in the Mediterranean Neogene: A consequence of geokinematic and climatic events. In: D. J. Starley and F. Wezel (Eds.), *Geological Evolution of the Mediterranean* (New York: Springer Verlag), pp. 560–570.

Tague, R., and Lovejoy, C. O. 1986. The obstetric pelvis of A.L. 288-1 (Lucy). *Journal of Human Evolution 15*: 237–255.

Taylor, C. R. and Rowntree, V. J. 1973. Running on two or four legs: Which consumes more energy? *Science 179*: 186–187.

Teleki, G. 1981. The omnivorous diet and eclectic feeding habits of chimpanzees in Gombe National Park, Tanzania. In: R. S. O. Harding and G. Teleki (Eds.), *Omnivorous Primates: Gathering and Hunting in Human Evolution* (New York: Columbia Univ. Press), pp. 303–343.

Tutin, C. E. G. 1979. Mating patterns and reproductive strategies in a community of wild chimpanzees (*Pan troglodytes schweinfurthii*). *Behavioral Ecology and Sociobiology 6*: 29–38.

Tutin, C. E. G. 1980. Reproductive behavior of wild chimpanzees in the Gombe National Park, Tanzania. *Journal of Reproduction and Fertility, Supplement, 28*: 43–57.

Ward, S., and Brown, B. 1986. The facial skeleton of *S. indicus.* In: D. Swindler and J. Erwin (Eds.), *Comparative Primate Biology, Vol. I, Systematics, Evolution, and Anatomy* (New York: Alan R. Liss), pp. 413–452.

White, T. D., Johanson, D. C., and Kimbel, W. H. 1981. *Australopithecus africanus*: Its phyletic position reconsidered. *South African Journal of Science 77*: 445–470.

CHAPTER 2

THE ORIGIN OF THE GENUS *HOMO*

■

Alan Walker*

■

INTRODUCTION

Anyone interested in their own roots should be interested in the origin of our own genus. For those researchers who believe that there has only ever been one species of the genus *Homo*, this origin was also the beginning of our own species. But others think that the situation was more complicated about 2 million years ago and that there were more than one species of early *Homo* then, that the genus had undergone a small radiation that produced several species. Although we know a great deal about the first undisputed members of our own genus that appear about 1.7 million years ago, there are still disagreements about which sort of hominid preceded them. We have only a few skulls, isolated bones, and two poorly preserved partial skeletons from about 2.5 to 1.8 million years ago. Discovering well-preserved early hominids from this time period is a critical priority.

The candidates for ancestral forms include species of the genus *Australopithecus*. These are known only from Africa and have their own specializations unlike those observed in *Homo*. Two species were specialized in having huge molars and **premolars**, and massive jaws. All species of *Australopithecus* were small-brained, upright, bipedal hominids with relatively big teeth, or **megadontia**, and significant body-weight dimorphism between the sexes. The earlier, smaller-toothed species are usually those presented as candidates for our own ancestors, but even those species with extreme specializations of the chewing apparatus have been considered by some researchers to be reasonable forerunners.

Specimens from East and South Africa about 1.7 million years old have been called *Homo erectus* (Figure 2.1). They show many of the hallmarks of humankind, including relatively large brain cases, completely modern limb proportions, and relatively small teeth. They still retain some primitive features such as strong brow

*Department of Cell Biology and Anatomy, Johns Hopkins School of Medicine, 725 North Wolfe Street, Baltimore, MD 21205

FIGURE 2.1

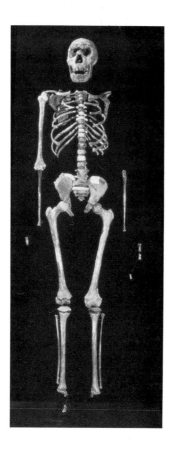

The mounted skeleton of early African *Homo erectus*. This specimen, KNM-WT 15000, is the most complete skeleton ever discovered of a fossil hominid. It was found near the western border of Lake Turkana in northern Kenya, at a site called Nariokotome III. The completeness and excellent preservation of the fossil provide a wealth of information about the individual, and the species it represents, as reported in detail in this chapter. Photo by R. I. M. Campbell, reproduced courtesy of the Kenya National Museum.

ridges, a very narrow pelvis, and long necks to their **femurs.** They were, as far as our small sample can show us, remarkably tall people. We have also been able to deduce that they already had the adaptation of having secondarily **altricial** babies, a uniquely human trait. This picture of our early ancestors is becoming clearer, and it allows us to examine the tempo and mode of evolution in our own lineage.

THE SPECIES OF *AUSTRALOPITHECUS*

There are four widely accepted species of the genus *Australopithecus.* Two of them, *A. afarensis* and *A. africanus,* are known from early in the record and are relatively unspecialized when compared with other *Australopithecus* or later *Homo.* The other two, *A. robustus* and *A. boisei* (sometimes known as *Paranthropus*), have specialized jaws and teeth and are found in later deposits. They overlap in time with *Homo erectus* in South and East Africa, respectively.

The earliest of these species, *A. afarensis,* is very well known from East Africa, especially by the good samples that have been found in Hadar, Ethiopia (see comprehensive descriptions of this material by D. C. Johanson and his colleagues in the *American Journal of Physical Anthropology*, vol. 57, no. 4, 1982). Some additional fragments have been found in Kenya, and a small sample of mostly jaws and teeth comes from the **type locality,** Laetoli in Tanzania (White, 1977). Claims have been made that other, fragmentary fossils should be placed in this species (Hill and Ward, 1988; Fleagle et al., 1991), but some are so fragmentary that it is not clear to me that they are necessarily hominid, and others are not especially convincing because their similarities with *A. afarensis* might be based on only primitive features. The sediments at Laetoli are about 3.6 million years old (Hay, 1987), but the better samples from Hadar are between about 3 and 3.3 million years old (Sarna-Wojcicki et al., 1985). Among the characteristics of these two species are postcanine megadontia, very large anterior teeth relative to the molars, small **endocranial volume** to body-weight ratio, and marked degree of body-weight dimorphism between the sexes. They had limb proportions that were neither like those of extant apes nor modern humans, and they had wide upper rims of the pelvic girdle. These characters are also found in *A. africanus* from South Africa. One or the other of these two species is considered most likely to be ancestral to later humans. Some workers think that the South African species shows features that are already specialized towards the South African "robust" lineage (Johanson and White, 1979), while others think that they are in our lineage and should be put in our genus (Robinson, 1972).

The characteristics of the South and East African species with specialized jaws and teeth, *A. robustus* and *A. boisei,* are well known for some parts of the skeleton and not at all for others. For recent summaries of our understanding of these species, consult the individual contributions in the book edited by Grine (1988). I do not consider that either of these species played a role in the origin of the genus *Homo,* but some workers consider them as candidates (Bromage, 1992).

■
HOMO HABILIS

Which Fossils Belong to This Species?

The species name ***Homo habilis*** was created by Leakey, Tobias, and Napier in 1964 to accommodate some fossils from Olduvai Gorge, Tanzania. Tobias (1991) has just published his mammoth work on these fossils which gives the most complete overview. Despite the number of words published on this species, however, there is not as much bony evidence as we would like. This is still an area of disagreement, with some people, like Professor Tobias and Johanson (1989), accepting this species as a single variable one that links *Australopithecus africanus* with *Homo erectus.* Others think that there are two or more species mixed up into *Homo habilis* because the morphological and size variation seems too great for only one species (Rightmire, 1990; Stringer, 1986; Walker and Leakey, 1978; Wood, 1991, 1992). This is an important disagreement because the implications of each viewpoint for the origin of our own genus are very different. In one case our genus arose rapidly with great morphological and behavioral changes about 1.8 million years ago, and in the other the transition was probably more gradual and may have begun much earlier. We return to this question at the end of this chapter. It is much easier to say what early *Homo erectus* was like than what *Homo habilis* was like and, with the evidence we have in hand, looking backward in time from *Homo erectus* is more profitable at the moment than looking forward in time from *Australopithecus.*

In contrast to the situation with the fossils of early *Homo erectus*, where apart from taxonomic niceties, most people agree that we are dealing with the early part of our clade, the situation with *Homo habilis* is a complex one. The size of the problem can be gauged by the realization that different paleontologists' **hypodigms** are drastically different. And this is not a matter of some fragmentary fossils that are difficult to agree on. Whole **crania** are placed by different people in different species or even genera.

The **type specimen** of *Homo habilis*, OH 7,* is a broken, subadult mandible with all teeth in place except the unerupted third molars. There are also some skull fragments that are probably associated with the jaw and some postcranial pieces that are possibly associated with it. Originally there were several additional specimens from Beds I and II at Olduvai, but over the years there has been a general acceptance of *Homo habilis* as a biologically valid species, and many more specimens have been attributed to it (Tobias, 1989a, 1989b). As the numbers of specimens increased, so too did the range of variation. It is this increased variation that is the root cause of the *Homo habilis* problem.

Some of the specimens placed in *Homo habilis* are characterized by being of very small size, with small cranial capacities and having body proportions that are like those of australopithecines or even in one case, limb proportions like chimpanzees (Johanson et al., 1987). Other individuals appear to have been much larger and to have isolated bones that look rather like those of early *Homo erectus*. A pelvic bone from East Turkana, for instance, looks remarkably like those of early *Homo erectus* (Rose, 1984), and femurs from about the same stratigraphic levels have been allocated, erroneously, to *Homo erectus* (Trinkaus, 1984). Some have even suggested that the variation seen in the sample from East Turkana is different from that seen in the sample from Olduvai, and that this geographic variation is a cause of confusion (Chamberlain, 1989). Others have suggested that we are dealing with more than one non-*boisei* species (Leakey and Walker, 1978; Stringer, 1986; Wood, 1992) and, further, that some bones allocated to *Homo habilis* are really those of *A. boisei* (Wood, 1976; Grausz et al., 1989).

Partial Skeletons of *Homo habilis*

Two specimens are of particular importance in the discussion about *Homo habilis* because they are of associated material and because one has been the subject of much recent publicity. The first of these, KNM-ER 3735 (Figure 2.2), consists of many cranial fragments and bits of postcranial bones, some of which have joint surfaces (Leakey and Walker, 1985). The initial piece was collected from East Turkana in 1975 and subsequent pieces were found by sieving at the site. It is dated between 1.88 and 1.91 million years (Feibel et al., 1989). A brief analysis showed that this individual weighed about 40 kg, had forelimbs built rather like those of modern chimpanzees, with strong, grasping fingers, and yet was bipedal (Leakey et al., 1989). Comparisons of the badly fragmented cranial pieces with other, more complete, crania from East Turkana shows that KNM-ER 3735 had a skull that was rather like that of smaller-brained specimens such as KNM-ER 1813, rather than larger-brained ones like KNM-ER 1470 (Wood, 1991).

The second important partial skeleton attributed to *Homo habilis* is OH 62, which was announced with great fanfare in 1987 (Johanson et al., 1987). It comes from well-dated **horizons** at Olduvai and must be about 1.8 million years old. It consists of fragmented cranial parts, the biggest of which is an upper jaw with three very worn teeth glued in place. Apart from these fragments, there is a shaft of a **humerus**, parts of an **ulna** and **radius**, and an upper end of a femur. Only the ulna

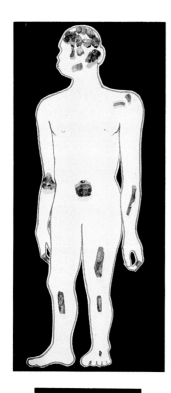

FIGURE 2.2

The fragments of a partial skeleton of *Homo habilis,* KNM-ER 3735, placed on a cartoon outline. This is one of only two very fragmentary skeletons of this species. Visible in the illustration are portions of the cranium, left scapula and radius, right humerus, right and left phalanges (finger bones), sacrum, right tibia, left femur, tibia, and metatarsal. Photo by A. Walker.

---

*Museum designations used in this chapter include specimens from Kenya and Tanzania. Hominid fossils from Olduvai Gorge, Tanzania, are numbered sequentially in order of their discovery after the prefix OH (for Olduvai Hominid). Specimens from the east side of Lake Turkana, Kenya, are designated with the prefix KNM-ER, for Kenya National Museum—East Rudolph (now East Turkana), and likewise, specimens from the west side of the lake are prefixed with KNM-WT, for West Turkana.

piece has a bit of **articular surface** preserved. The describers compared it with the much better preserved partial skeleton of *A. afarensis* known as "Lucy" from Hadar, Ethiopia. This useful exercise resulted in the following conclusions. The Olduvai individual was tiny as an adult and was as small or smaller than Lucy. The cranial fragments show that it was an old individual with a skull of the same size and shape as specimens often allocated to *Homo habilis* (OH 13 and KNM-ER 1813). The entire forelimb must have been longer and the femur less robust than in the small individual of *A. afarensis*. The estimated ratio of humerus length to femur length is about 0.95, close to that of chimpanzees, although there is no doubt that this individual was an habitual, upright biped.

Hartwig-Scherer and Martin (1991) also looked at this specimen. They confirmed the low body weight estimate (25 kg, about 20% less than Lucy), and they also concluded that in the parts they could measure the skeleton showed more similarity with African apes than with the fossil from Hadar. And, they said, if OH 62 is a hominid it could not fit comfortably as an ancestor of *Homo erectus* and if it belonged in *Homo habilis* then that species should be excluded from the human lineage.

These two very fragmentary partial skeletons differ in size by the amount that is expected if they were male and female of a sexually dimorphic primate species such as the chimpanzee (Leakey et al., 1989). But there are also isolated postcranial fossils from the same time period from much larger individuals. This greatly increases the range of size variation in non-*boisei* hominids from this time period and makes it unlikely, in my view, that we have sampled only one species. Wood (1992) has compiled the latest review of some of these issues and concludes that there were three non-*boisei* species about 2 million years ago. If he is correct, these would have the names *Homo habilis, Homo rudolfensis* and *Homo ergaster*. For Wood, the first includes OH 62 and KNM-ER 1813, the second KNM-ER 1470, and the last KNM-ER 3733 and KNM-WT 15000. The first two sets of fossils do come from the same time periods, but I see only fragmentary evidence for the last species (or early *Homo erectus* in my terms) earlier than 1.7 million years ago.

EARLY *HOMO ERECTUS* IN AFRICA

Early Discoveries

Fragmentary early hominids unlike contemporaneous *Australopithecus* have been known for some time from South and East Africa. As is usually the case with fragmentary fossils, these were often the subject of disputes (Figure 2.3). When the first of these fossils were found at the limestone cave site of Swartkrans, South Africa, which had previously yielded remains only of *A. robustus*, the several pieces of jaw were given a new genus and species name, *Telanthropus capensis*, but were later placed in *Homo erectus* (Robinson, 1961). Not all authorities accepted that these pieces were anything other than small individuals of *Australopithecus*. Some workers continued to believe until the mid-1970s in the single species hypothesis which postulated that there was only one hominid species at any one time in the past (Wolpoff, 1971). The use of the name *Homo erectus* for these early non-*Australopithecus* hominids is still debated and the reasons are complex. We cannot reach a consensus even now that we have a nearly complete skeleton of this type of human. The reasons are partly historical and partly to do with new methods of classification.

I am not concerned here with whether the name *Homo erectus* is really the appropriate one, but the literature can be confusing if this point is not made clear. The

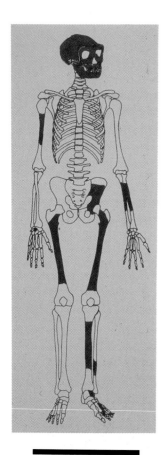

FIGURE 2.3

Known parts of *Homo erectus* from all areas of the world in 1976. Many of the most complete specimens at that time were from Asia, including nearly complete crania from the cave site of Zhoukoudien, China, and a nearly complete femur from Java. The African material known before 1976 was extremely fragmentary and difficult to interpret.

term *Homo erectus* is strictly applied to fossils found more than a century ago by Eugene Dubois in deposits about 1 million years old in Java. Fossils of similar type found in China were first given their own names, but now are considered to be the same species as the Javan hominids. This species is also now thought to be the antecedent to our own by practically all authorities. A few still believe, however, that the Far Eastern representatives are so specialized that they must represent endemic populations that were not ancestral to any living ones (Andrews, 1984).

The first good evidence of this species from Africa came when Louis Leakey found OH 9. This massive **calvaria** is about 1.2 million years old and resembles those from the Far East quite closely. Ironically, although this fossil was the only one that Leakey found himself at Olduvai, he was so convinced that *Homo erectus* was a sidebranch of evolution and that humans proper arose in Africa, that he denied his fossil a place as a representative of our ancestors (Leakey, 1964). As we shall see, it is still difficult to gather a consensus about the correct name for these fossils even when good evidence of this type of human is now known from earlier deposits in East Africa.

Richard Leakey began expeditions to the east side of Lake Turkana (then still called Lake Rudolf) in 1968. In what was almost a replay of the events at Swartkrans, the first hominids found were megadont *Australopithecus*. Some small, fragmentary specimens were also discovered, but these were, again, not complete enough to refute the single species hypothesis. They were placed by their describers in the genus *Homo*, but no species name was given. Meave Leakey found a fragmentary partial skeleton of one of these early hominids in 1971, but because postcranial bones of *Homo erectus* were very poorly known from anywhere, it was not recognized as such. Kamoya Kimeu found a much more complete skeleton in 1973. The first pieces were cranial and dental, but it was noticed that these occurred in a place in which fossil bone was scattered over an area about the size of a football field. This area was scraped down to bedrock and sieved. Thousands of fragments of mammal bones and bone fragments were collected in this way and the hominid ones removed. These were glued together to make many parts of a single skeleton. Bones could be sorted easily because only the hominid bones had distinctive surfaces that were caused by disease. The adult individual had suffered a bout of **periostitis** shortly before death and this had resulted in nearly all the bones being covered with an external layer of rapidly deposited **woven bone**. The diseased bone made the study of normal morphology very difficult, but the possible diseases that might have been involved did give us reasons to think about the life style of this early hominid (Walker et al., 1981). We recognized that this might have been caused by a disease no longer manifesting itself, or one that gave different symptoms then. Of diseases of modern people, **hypervitaminosis A** results in bony changes that were similar to those seen in the fossil. The most likely source of excess vitamin A, we thought, would have been the consumption of adult carnivore livers. This seemed reasonable to us because it had long been known that these early hominids were gathering bones with flesh on them, which would have brought the hominids into close contact with other carnivorous species. Skinner (1991) has suggested that bee brood (eggs, pupae, and larvae) might be the source of excess vitamin A rather than liver. This is possible, but since all modern people use smoke to control bees and since the evidence for the origins of the early use of fire by people is weak, I regard this alternative as less likely.

The Discovery of the Nariokotome Hominid

In August of 1984, Kamoya Kimeu found a small cranial piece of *Homo erectus* at the 1.5 million year old site of Nariokotome III on the west side of Lake Turkana

FIGURE 2.4

Map showing the localities of the Nariokotome III site near the western shore of Lake Turkana. Koobi Fora on the east side of the lake, and the Omo River on the north, are also areas that have yielded fossil hominids. The Kenya-Ethiopia border traverses the northern part of the lake.

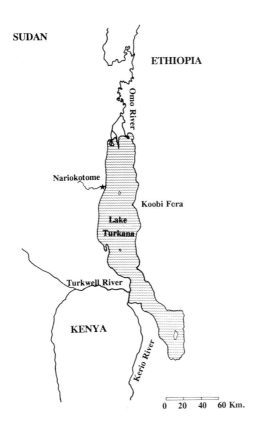

(Figure 2.4). This was the first piece of a specimen that would prove to be the most complete early hominid skeleton ever found, since designated KNM-WT 15000. This skeleton has already provided a wealth of new information about early *Homo erectus* and has given us insights into life history, growth and development, and evolution that hundreds of less complete specimens never did. This single specimen enabled us to distinguish unequivocally between most of the bones of *A. boisei* and *Homo erectus,* a task that was more difficult than an outsider might imagine. It also allowed us to reexamine the more fragmentary fossils from previous collections and make more use of them. This specimen has become equivalent to a *keystone species*, in the jargon of ecology, because the interpretation of so many others is dependent on it.

Drs. Craig Feibel and Frank Brown (1993) carried out detailed studies of the stratigraphy and microfauna at the excavation site (Figure 2.5). The insights from their study, together with **taphonomic** studies of the mammal bones and their distribution give us a fairly complete picture of what happened to this individual *Homo erectus* from his death until his skeleton was excavated. We do not know the cause of the boy's death. There are no signs of trauma or toothmarks that might be due to his being killed by a predator. The only sign of disease is a small **periodontal abscess** on the right side of the mandible. We can see that this was associated with the shedding of the last **milk molar** and the concomitant eruption of the second premolar. Two small indentations within this pocket of abscessed bone show that tiny pieces of milk molar root were left in the jaw when the milk tooth was shed. Whether this abscess led to an infection that killed the boy is impossible to tell. In any case, the body came to lie in a very shallow swamp.

This swamp was situated immediately to the west of the main proto-Omo River which at this time was flowing in a huge braided stream over the area where the present Lake Turkana lies. The swampy areas were low-lying patches on a wide and

FIGURE 2.5

The site of Nariokotome III in 1986 during excavation of KNM-WT 15000. Notice the sparse vegetation and the erosion caused by downcutting of a tributary to Lake Turkana that exposed fragments of the skeleton on the surface. Photo by A. Walker.

flat floodplain rich in seasonal grasses and large animal life. The floodplain was inundated several times a year, and the depressions received fine sediment during flooding. Thus the positions of swamps changed as older ones filled with sediment and new ones got established in the next flood. These swamps attracted large and small animals. Three-toed horses, pigs, antelopes, hippopotamuses, wading birds, and monitor lizards left their bones with those of the youthful *Homo erectus* in the Nariokotome swamp. The bed of the swamp was an ashy silt that had been trampled by the feet of large mammals. The numerous root casts that we found over the swamp bed attest to its being filled with swamp grasses, sedges, or reeds. The animal bones were distributed all over the area we excavated, while the hominid skeleton was concentrated mostly in the southwest corner of the site. I measured the orientations of the bones before they were lifted from the rock and compared them with the results of experiments carried out on the way bones are oriented by water flow. The bones on the site were aligned in two main directions, roughly east-west and north-south. This pattern of two alignments perpendicular to each other is typical of bones lying in very shallow water that is either running or has a current caused by the wind. That the swamp was very shallow is confirmed by other evidence from the site. We found many swamp snail opercula—the protective lid that closes the opening when a snail retracts into its shell—when we were excavating the hominid-bearing layer of rock. This snail is an air-breathing snail that can only live in shallow water and ascends plant stems to breathe. One of the commonest microfaunal elements is a freshwater sponge that cannot live on muddy substrates, but needs plant stems to which it attaches. The sponge confirms that plant stems were plentiful in the swamp, and the snail shows that the stems emerged from the surface of the water.

The concentration of hominid bones allowed us to reconstruct the rotting and breakup of the body in some detail. Paleolithic archeologists frequently use a refitting diagram to show how flakes of stone tools that were once part of a single stone become scattered over a site. I used the same approach on the hominid skeleton—a sort of "neck bone attaches to the head bone" approach. My diagram showed that the body must first have been lying face down in the water a little to the north of its final position. Some of the upper and lower front teeth slipped out and fell in one place as the gum tissues rotted. The corpse gradually fell apart as the flesh rotted further and was eaten by scavenging catfish. But the main agency of hominid bone dispersal was the trampling and kicking of large animals as they waded through the swamp. Several of the bones were broken by trampling. The weak current from the north together with trampling ensured that the bones came to rest in the mud toward the edge of the swamp, with the upper-body elements generally to the west and the lower-body elements toward the east. The next season's flooding covered the bones with silt. Later a thick ash washed over the area and the bones remained undisturbed until a small tributary of the Nariokotome River cut into the ancient swamp deposits in the last few hundred years. The roots of modern *Acacia* and *Salvadora* trees had found their way through the sediments and caused some root damage to a few of the bones and teeth. The backcutting of the Nariokotome tributary had already exposed some of the bones by the time Kamoya Kimeu found the first pieces, and we now think that many parts of the skeleton, including most of the small hand and foot bones and smaller **epiphyses** were exposed and have been lost in the modern sand river (Figure 2.6).

The Age and Size of the Nariokotome Hominid

The age of the individual was determined initially by tooth eruption stage to about 12 years in human terms (Brown et al., 1985). The upper **deciduous** canines were still in place at the time of death, although their roots had begun to **resorb** in preparation for the eruption of the permanent canines. Dr. Holly Smith (1993) has since determined that the age (in modern human terms) was 11 ± 1 years. It is also possible to determine the age of a skeleton based on when the ends of the limb bones, the epiphyses, fuse to the bone shafts, thereby terminating further elongation of that bone. She points out that in terms of epiphyseal union the individual was about 13 years old and in terms of stature about 15 years old. When we had initially determined the stature we used equations based on adult individuals from North America (Brown et al., 1985). We knew, of course, that we were dealing with a subadult, but thought that this would not greatly affect our stature estimate. We were interested then in getting a rough idea of overall body size because it was clear to us on the site that this individual was tall. New stature estimates based on limb segment lengths show that the youth was about 1.6 m (5'3") at death and, had he lived to maturity and continued growing like modern people, would have been over 6' tall (Ruff and Walker, 1993). When we first saw the size of his limb bones in 1984 we thought that we had excavated an individual *Homo erectus* from the very tall end of the size range. But we were looking at the first complete limb bones known of early African *Homo erectus*, and when we compared them with the more fragmentary specimens from earlier collections we found that all these hominids were tall. For those six individual early *Homo erectus* for whom we could accurately determine stature and body weight, the mean adult stature was 1.7 m (5'7") and body weight was 58.1 kg (Ruff and Walker, 1993).

Two questions immediately come to mind. First, why were these people so tall—or why are many modern populations so short? Second, what selectional change led to the size increase in the antecedents to early *Homo erectus*? The answer to the first question seems to be that many modern populations have short people because they are small in mass and that this has to do with the quite recent adoption of agriculture. Evidence from many parts of the world shows that taking up agriculture leads to a decline in the health of individuals within a population (Cohen and Armelagos, 1984) and that selection for small body size is a consequence of the uncertainty of resources. This is similar to the effect seen on many islands where resources can become limited for animals of large body size. The answer to the second question seems to be related to a dietary change, where the consumption of high-protein food

FIGURE 2.6

Kamoya Kimeu, who found the first fragments of the Nariokotome skeleton, scans the exposed desert surface near Lake Turkana. Initial fossil discoveries are often made by explorations of eroded land surface; usually excavation only becomes productive after surface prospecting locates the richest areas or the ones holding important specimens. Photo by A. Walker.

FIGURE 2.7

FIGURE 2.7

Lateral view (left) and frontal
view (right) of KNM-WT 15000.
The well-preserved cranium
provides information about the
age, size, and cranial capacity of
the individual, and by comparison
to other fossil finds, it also
contributes to information about
population variability. Photo by
A. Walker, reproduced courtesy of
the Kenya National Museum.

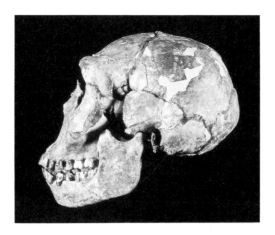

such as meat can enable bigger body size to be selected for following a behavioral
change. The archeological record shows that early hominids consumed some meat,
probably scavenged, before the time of early African *Homo erectus*, but that
changes in the stone tool technology and the evidence of butchered animal bones
show that active hunting and meat eating was taking place regularly only with the
advent of *Homo erectus* (Shipman, 1986). Increased body size would be an advantage in competition for food with other large predators.

The Morphology of the Nariokotome Hominid

I can only sketch here the wealth of morphological information to be found in this
skeleton (Figures 2.1 and 2.7). But more than morphological information can be
gained. We can also begin to deduce when certain behavioral or physiological human
traits evolved. Among these are the following:

1. Had *Homo erectus* developed the uniquely human trait of meeting heat stress
   with a sweating, rather than the typical primate panting and sweating strategy?
2. Had *Homo erectus* developed the uniquely human trait of having secondarily altricial infants?
3. Did *Homo erectus* have spoken language?
4. Were these early humans mostly right handed?

Most of the morphology of this individual can be matched in modern human
skeletons, but certain things are different. Some of these are retentions of primitive
conditions and others are more puzzling. The proportions of the body are extremely
like those of certain equatorial modern humans. We have noted the boy's body
shape, which seems to conform to *Allen's rule*. This rule is that people and other
mammalian species near the geographic poles are short and thickset with short **distal**
segments to their limbs, whereas people living at the equator are tall and very slender
with long distal limb segments. The latter was the case for the Nariokotome skeleton
and therefore suggests that Allen's rule had been in effect for that population 1.5
million years ago. The adaptive basis for the rule is that the ratio of surface area to
body volume is maximized in an animal that sweats under heat stress, and it is mini-
mized to conserve body heat in animals evolving under cold conditions. Following
Allen, Ruff (1991) has now developed a climatic model of human body proportions
and has shown that the Nariokotome youth was hypertropical in body build, being

tall and very slim, like modern Dinka and Turkana people. Trinkaus (1981) used data from modern populations to examine the relationship between limb proportions and mean annual temperature. He developed regression equations from which we can predict the mean annual temperature of the Nariokotome region when the youth was alive. The upper limb gives an estimate of 30.8°C and the lower limb 29.2°C. These are consistent results and give distinctly equatorial mean annual temperature predictions. In a paleontological reconstruction of the climate, Feibel et al. (1991) reported that the Lake Turkana region has had the same climate as today since about 1.5 million years ago. This, together with the evidence from the skeleton, points to these populations of *Homo erectus* being adapted by sweating for efficient cooling under high ambient temperatures. This means that humans' extraordinary ability to sweat (Stoddart, 1990) has its origins before this time and also that these people must have had to live near water sources.

We have determined the cranial capacity at 880 cm³ (Begun and Walker, 1993). Modern humans have about 95% of their adult brain weight by the age of 10 years, so allowing for a little increase and following Tobias (1971), we calculate an adult cranial capacity of 909 cm³. This is about on the mean for *Homo erectus* from Java (Rightmire, 1990) and suggests that not much change in cranial capacity took place from 1.5 to 0.7 million years ago. But this figure alone is not sufficient for a discussion about brain size because relative brain size is the most important factor. When allowances are made for body size, it seems that early *Homo erectus* individuals had a relative brain size like those of their putative ancestors, *Homo habilis* (Walker, 1991). But note that this is only when those individuals of *Homo habilis* with large brains are analyzed together with body weights from the largest individuals. The asymmetries seen in the cranial **endocast** are similar to those reported for right-handed male humans (Begun and Walker, 1993).

The teeth of the Nariokotome youth are very close in size, shape, and tiny details of morphology to those of *Homo erectus* from China (Brown and Walker, 1993; Weidenreich, 1937). This includes **shovel-shaped** central and lateral upper incisors. Shoveling of incisors is often cited as a trait of modern Far Eastern and native American populations, and its presence in early Chinese populations had been taken to show genetic continuity there (Weidenreich, 1937). But it now appears more likely that this is a primitive trait for the genus *Homo* and as such cannot be used to argue that this is a special Far Eastern trait.

Although the Nariokotome youth was only about 11 years old in human terms, his facial development had progressed to the point of his having quite large maxillary and frontal air sinuses. Compared with an adult female cranium of early African *Homo erectus*, KNM-ER 3733, the facial skeleton was already larger and more robust (Figure 2.8). It is not beyond the bounds of possibility that had this individual lived he would have grown up to have a cranium as large and robust as OH 9 (Rightmire, 1990). The cranium had not developed the strong **supraorbital** and **occipital tori** or the thick brain case that are so typical of *Homo erectus,* but there are some hints in the youthful cranial structure that these would have grown had he lived.

The **vertebral column** shows the normal pattern of curvature found in modern humans, with the upper part of the back convex and the lower part concave when viewed from behind. There are three surprises in the **vertebrae.** The first is that the **neural canals**—the large central opening for passage of the spinal cord—are small, particularly in the **thoracic,** or upper back, region. The second is that the **neural spines**—the bony blades that extend dorsally from a vertebra to serve as attachments for ligament and muscle—are longer and more erect than is expected for a modern human. The third surprise is that there were six **lumbar,** or lower-back vertebrae, one more than in modern humans. Taking these points in order, the neural canal has been studied by Dr. Ann MacLarnon (1993). She found that it was very like a typical hu-

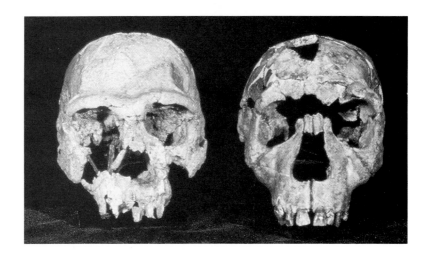

FIGURE 2.8

Frontal views of a fully adult female specimen of *Homo erectus* (KNM-ER 3733, left) and a subadult male (KNM-WT 15000) showing the degree of sexual dimorphism. For such a young individual, KNM-WT 15000 is already fairly large, with pronounced temporal lines (crests for chewing muscles on the side of the head), and parasagittal flattenings (on either side of the skull midline) that suggest stronger cranial crests probably would have developed in the adult male. Photo by A. Walker, reproduced courtesy of the Kenya National Museum.

man canal, but that it lacked the unique human thoracic enlargement (Figure 2.9). The human enlargement of the canal is correlated with an increase in the amount of gray matter in the spinal cord. Gray matter contains nerve cell bodies as well as nerve fibers. Its abundance in the cord at the thoracic region indicates motor activity in the **intercostal** muscles, which create movements of the rib cage, or in the anterior abdominal wall musculature. MacLarnon gives two reasons why this might be so. The first is that the enlargement is related to increased control over these muscles for bipedal walking and running. This is not likely, since hominids had been bipeds for 2.5 million years before this individual came on the scene. The second reason is for the fine muscular control of the breathing apparatus needed in human speech. This was not developed in australopithecines because these hominids did not have spoken language. This point is discussed subsequently. The more erect neural spines have been explained by Latimer and Ward (1993) to be an adaptive response related to the increased dorsal curvature seen in the ribs of this individual over the more normal human condition. That this individual had six lumbar vertebrae is interesting in that it probably represents the original early hominid condition. Old World monkeys and early, fossil apes have six or seven lumbar vertebrae. Great apes have three or four lumbar vertebrae and humans usually have five, but the only other early hominid in which this can be checked had six (Robinson, 1972). Thus either lumbar elongation and flexibility evolved anew after the ape-hominid split — an evolutionary reversal — or else humans have retained a primitive elongated lumbar column while the living Asian and African great apes independently evolved a shorter one — a case of parallel evolution.

In the upper limb the bones are remarkably like those of modern humans, and there are even signs of asymmetries related to right handedness. This finding corrob-

FIGURE 2.9

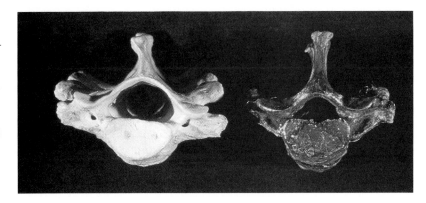

FIGURE 2.9

Superior views of the lowest cervical (neck) vertebra of a modern adult human (left) and of KNM-WT 15000 (right). Notice the much smaller diameter of the neural canal and the longer neural spine in the fossil specimen. Photo by A. Walker.

orates what we found in the brain asymmetries. Toth (1985) has shown, too, that the stone-tool makers of this time were right handed, so that right handedness appears to date back to at least 1.5 million years.

The **pelvis** can be reconstructed with some accuracy, despite the juvenile state of nonunion of the three main bones on each side and the unfused **sacrum.** The pelvis is narrow, both externally and in the pelvic inlet. The narrow diameter across the upper blades of the pelvic girdle (the bi-iliac diameter) fits with Ruff's climatic model of an individual who was adapted to an active life style on the open grasslands of equatorial Africa. The narrow pelvic inlet is important in determining whether early *Homo erectus* had the uniquely human trait of having secondarily altricial infants. This peculiar trait has been discussed by Gould (1977) and Martin (1983) and is of great importance, I feel, in predicting certain aspects of early human behavior. Humans are the only primates to extend the rapid intrauterine growth rate of the fetal brain into the postnatal period. All other primates studied so far discontinue the rapid rate at or about birth, whereas humans continue the fast rate of growth for about a year after birth and only then slow it down (Figure 2.10). This leads to the altriciality of human infants, but this is only altriciality of the infant motor skills and not sensory ones. This is why it has been called secondary altriciality. If early African *Homo erectus* were like other primate species, then infants would be born with half of the adult brain weight, rather than one-quarter to one-third as in modern humans (Trevathan, 1987). If this were the case, they would be born with as large a brain as the full-term fetus of modern *Homo sapiens*, because their adult capacities are over 800 cm³. Ruff and Walker (1993) have estimated the birth canal dimensions in *Homo erectus* based on the Nariokotome youth and other specimens. The critical canal dimensions are all much smaller than those of modern women. We estimate a head size for the full-term infant equivalent to only a 32- to 33-week human fetus, which would have a brain size of about 200 cm³. To reach an adult capacity of over 800 cm³, therefore, early *Homo erectus* infants must have had a pattern of relative brain growth like that of modern people, with substantial rapid growth for the first part of the infant's life. Martin (1983) thought that until about 1.5 million years ago early hominids could maintain the normal primate pattern of growth by making adjustments in the rate and/or extent of fetal growth. We now see that this seems not to have been so.

There are two not necessarily exclusive, adaptive reasons why early hominids changed to having secondarily altricial infants. Either the narrowness of the body as an adaptation for increasing the surface area to volume ratio could not be compromised, or the biomechanical efficiency of a small distance between the hip joints (interacetabular distance) and long femoral necks could not be. The femurs of *Homo erectus* show the long necks that have been reported for the genus *Australopithecus*

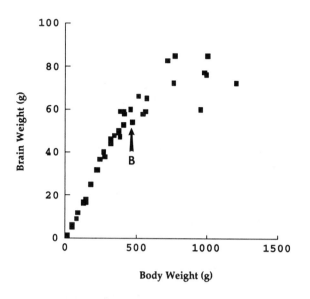
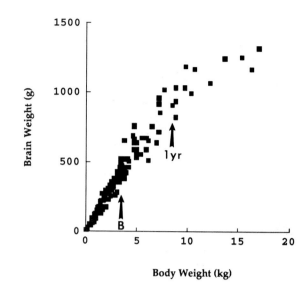

FIGURE 2.10

The growth of the brain and body in macaque monkeys (left) and humans (right) from conception until several months after birth. Body size is plotted on the horizontal axis, while corresponding increases in brain size are shown on the vertical axis. B marks the point of birth. Notice that in macaques the curve is steep until that point. After birth, the fast rate of brain growth slows down. In humans, note that unlike the macaque, the brain continues to grow at the same fast fetal rate until about 1 year after birth. Only then does the growth rate slow. Data from Martin (1983).

and which, although not seen in great apes, must be primitive for hominids themselves. The long necks mean that, even though the pelvis is narrow, the upper parts of the femur are widely spaced at the point where important muscle groups attach that are used in walking. This makes the **abductor** mechanism biomechanically efficient during bipedalism (Figure 2.11).

The answers to the four questions about behavior listed at the beginning of this section seem to be "yes." By showing that the Nariokotome youth was climatically adapted following Allen's rule, we also showed indirectly that its cooling strategy must have been the same as in humans, with a reliance on an enormous sweating organ (the skin). We have seen that **obstetrical** conditions and the size of the pelvis of *Homo erectus* showed that the adaptation of having infants that were secondarily altricial was already in place 1.5 million years ago. We showed that these early humans were probably mostly right handed and that the tools confirmed it. And finally, despite right handedness and the left cerebral lateralization associated with it, we have seen that the nervous control of the intercostal and abdominal wall muscles needed for articulate spoken language was probably not developed then. This last point has more significance that it seems at first glance. It used to be thought that certain **cortical** regions of the brain related to speech could be observed on endocranial casts of skulls. Tobias (1991) gives the most recent review of this field. Evidence for the development of some cortical speech areas was claimed for early hominids and Tobias himself believes that *Homo habilis* shows the first clear evidence of this. But work by Marcus Raichle and his colleagues at Washington University using positron emission scanning of normal subjects carrying out routine language tasks has changed this view. The dynamic method Raichle uses allows areas of brain activity to

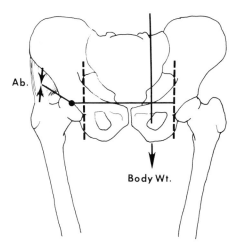

Ab.

Body Wt.

FIGURE 2.11

Diagram to show the abductor mechanism of the human hip. During walking and running the force of body weight × the distance from its line of action to the center of the hip joint must be balanced by the abductor muscle force × its distance from the center of the hip joint. Note that the line through which body weight acts is shifted to the side of the leg that is off the ground. Making the femoral neck longer and/or the distance between the hip joints shorter improves the mechanical advantage of the abductor muscles. But narrowing the distance between the hip joints decreases the diameter of the birth canal, shown in dashed lines.

be identified and monitored in active subjects. This work has shown that the traditionally defined motor speech area of Broca is probably more involved in coordination of movements than in speech. Cortical areas have been identified that are activated during speech, but these are either more anterior than **Broca's area** or are buried in **opercular cortex** that does not show on the endocasts of skulls. The endocranial evidence for language has become, therefore, much more insecure. This new development, together with MacLarnon's evidence for less trunk muscle control in *Homo erectus,* leads me to think that the origins of articulate speech were much later than 1.5 million years ago. This is in accordance with some new thinking concerning the lack of symbolic meaning in Paleolithic stone tools and the late origin of overt evidence of symbolic thought (Noble and Davidson, 1991).

THE ORIGINS OF *HOMO ERECTUS*

Most authorities agree that the early African members of the genus *Homo* such as the Nariokotome specimen are from the early part of the human lineage, even if they do not agree on the name for them. But the origins of these populations are still a matter of dispute. Most workers fall into two groups. The first group thinks that there was only one hominid species besides *A. boisei* at this time and that it contained the antecedent populations to early African *Homo erectus,* which would be called *Homo habilis.* The second group thinks that there were two non-*boisei* species at this time, *Homo habilis* and something else. The problem is not easy to solve because there are about as many hypodigms of *Homo habilis* as there are workers in the field. Even if we cannot solve this issue right now, we can give two alternative accounts based on these two perspectives.

If there was only one non-*boisei* hominid species antecedent to *Homo erectus* then the following can be said about the transition. *Homo habilis* in this account would be a highly variable, sexually dimorphic species with body-size dimorphism approaching that of species of *Australopithecus*. Females were short, perhaps only a meter tall, and had proportions of the humerus and femur more like those of chimpanzees than humans (Johanson et al., 1987; Johanson, 1989). They were relatively megadont and their skulls retained many features reminiscent of *Australopithecus* (Walker, 1976), including the flat margins of the **pyriform aperture** and nasal region. OH 62 would be an example of this species existing about 1.8 million years ago.

The transition to *Homo erectus* took place in a **punctuational** event, then, in which this *Australopithecus*-like species changed to become a larger one with less sexual dimorphism. An early example of this younger species would be the 1.7 million-year-old cranium, KNM-ER 3733. The females became nearly as large on average as the males in the preceding species. Brain size increased along with body size, and the limb proportions became essentially human. The transition also coincided with major changes in **lithic** technology. This hypothesis is put forward by the discoverers and describers of OH 62 (Johanson et al., 1987; Johanson, 1989).

The second account involves two hominids living alongside *A. boisei* about 2 million years ago. One of the two species that are presently lumped in *Homo habilis* was ancestral to *Homo erectus* and is represented by specimens such as KNM-ER 1470, 1471, and 3228. Morphologically these specimens are quite close to those of early *Homo erectus*. Ruff and Walker (1993) have estimated the stature and body weights of three of these early hominids. Early *Homo erectus* were about 16% taller and 26% heavier than those specimens. This species is not, however, well known. If this medium-sized species is antecedent to *Homo erectus* then there need not have been a punctuational event and, in fact, the genus might have originated much earlier, perhaps with the appearance of stone tools about 2.4 million years ago. The other, smaller species is better represented in both South and East Africa and includes specimens like OH 24 and 62, and KNM-ER 1813. This two-species hypothesis was first put forward by Leakey and Walker (1978).

The testing of these two competing hypotheses is critical to our understanding of the origin of the genus *Homo*. Testing them should determine if our lineage evolved in a punctuated evolutionary manner and if major external stimulants such as global climatic shifts affected the course of our own evolution. Whether we can recognize any correspondence between morphological changes and major faunal changes or the appearance of new lithic technologies depends on our ability to sort out the **phylogeny** of the earliest members of the genus *Homo*. We have been able in the past to find new evidence to test hypotheses (for example, Leakey and Walker, 1976), so it should be a matter of urgency that we plan field expeditions to look for the more complete fossils that will solve this problem.

■
──────

SUGGESTED READINGS

Aiello, L., and Dean, C. 1990. *An Introduction to Human Evolutionary Anatomy* (London: Academic Press).

Grine, F. E. 1988. *Evolutionary History of the Robust Australopithecines* (Chicago: Aldine).

Rightmire, P. G. 1990. *The Evolution of Homo erectus* (Cambridge: Cambridge Univ. Press).

Tobias, P. V. T. 1991. *Olduvai Gorge Volume 4, The Skulls, Endocasts and Teeth of Homo habilis* (Cambridge: Cambridge Univ. Press).

Walker, A., and Leakey, R. E., (Eds.) 1993. *The Nariokotome Homo erectus Skeleton* (Cambridge: Harvard Univ. Press).

REFERENCES

Andrews, P. 1984. An alternative interpretation of the characters used to define *Homo erectus*. *Courier Forschungsinstitut Senckenberg 69*: 167–175.

Begun, D., and Walker, A. 1993. The endocast of the Nariokotome hominid. In: A. Walker and R. E. Leakey (Eds.), *The Nariokotome Homo erectus Skeleton* (Cambridge: Harvard Univ. Press).

Bromage, T. 1992. Faces from the past. *New Scientist 133*: 38–41.

Brown, B., and Walker, A. 1993. The dentition of the Nariokotome hominid. In: A. Walker and R. E. Leakey (Eds.), *The Nariokotome Homo erectus Skeleton* (Cambridge: Harvard Univ. Press).

Brown, F., Harris, J., Leakey, R., and Walker, A. 1985. Early *Homo erectus* from west Lake Turkana, Kenya. *Nature 316*: 788–792.

Chamberlain, A. T. 1989. Variations within *Homo habilis*. In: G. Giacobini (Ed.), *Hominidae* (Milan: Jaca), pp. 175–181.

Cohen, M. N., and Armelagos, G. J. 1984. *Paleopathology at the Origins of Agriculture* (Orlando: Academic Press).

Feibel, C. S., and Brown, F. H. 1993. Microstratigraphy and paleoenvironments of the Nariokotome hominid site and associated strata. In: A. Walker and R. E. Leakey (Eds.), *The Nariokotome Homo erectus Skeleton* (Cambridge: Harvard Univ. Press).

Feibel, C. S., Brown, F. H., and McDougall, I. 1989. Stratigraphic context of fossil hominids from the Omo Group deposits: Northern Turkana Basin, Kenya and Ethiopia. *American Journal of Physical Anthropology 78*: 595–622.

Feibel, C. S., Harris, J. M., and Brown, F. 1991. Paleoenvironmental context of the late Neogene of the Turkana Basin. In: J. M. Harris (Ed.), *Koobi Fora Research Project, Volume 3* (Oxford: Clarendon).

Fleagle, J. G., Rasmussen, D. T., Yirga, S., Bown, T. M., and Grine, F. E. 1991. New hominid fossils from Fejej, southern Ethiopia. *Journal of Human Evolution 21*: 145–152.

Gould, S. J. 1977. *Ontogeny and Phylogeny* (Cambridge: Belknap).

Grausz, H. M., Leakey, R. E., Walker, A., and Ward, C. V. 1989. Associated cranial and post-cranial bones of *Australopithecus boisei*. In: F. E. Grine (Ed.), *Evolutionary History of the Robust Australopithecines* (Chicago: Aldine), pp. 127–132.

Grine, F. E. 1988. *Evolutionary History of the Robust Australopithecines* (Chicago: Aldine).

Hartwig-Scherer, S., and Martin, R. D. 1991. Was "Lucy" more human than her "child"? Observations on early hominid postcranial skeletons. *Journal of Human Evolution 21*: 439–449.

Hay, R. L. 1987. Geology of the Laetoli area. In: M. D. Leakey and J. M. Harris (Eds.), *Laetoli — A Pliocene Site in Northern Tanzania* (Oxford: Clarendon), pp. 23–47.

Hill, A., and Ward, S. 1988. Origin of the Hominidae: The record of African large hominoid evolution between 14 my and 4 my. *Yearbook of Physical Anthropology 31*: 49–83.

Hill, A., Ward, S., Dieno, A., Curtis, G., and Drake, R. 1992. Earliest *Homo. Nature 355*: 719–722.

Johanson, D. C. 1989. A partial *Homo habilis* skeleton from Olduvai Gorge, Tanzania: A summary of preliminary results. In: G. Giacobini (Ed.), *Hominidae* (Milan: Jaca), pp. 155–166.

Johanson, D. C., Masao, F. T., Eck, G. E., White, T. D., Walter, R. C., Kimbel, W. H., Asfaw, B., Ndessokia, P., and Suwa, G. 1987. New partial skeleton of *Homo habilis* from Olduvai Gorge, Tanzania. *Nature 327*: 205.

Johanson, D. C., and White, T. D. 1979. A systematic assessment of early African hominids. *Science 202*: 321–330.

Latimer, B. L., and Ward, C. V. 1993. The vertebral column of KNM-WT 15000. In: A. Walker and R. E. Leakey (Eds.), *The Nariokotome Homo erectus Skeleton* (Cambridge: Harvard Univ. Press).

Leakey, L. S. B. 1964. Very early African Hominidae and their ecological setting. In: F. C. Howell and F. Bourliere (Eds.), *African Ecology and Human Evolution* (London: Methuen), pp. 448–457.

Leakey, L. S. B., Tobias, P. V., and Napier, J. R. 1964. A new species of the genus *Homo* from Olduvai Gorge. *Nature 202*: 7–9.

Leakey, R. E., and Walker, A. 1976. *Australopithecus, Homo erectus* and the single species hypothesis. *Nature 261*: 572–574.

Leakey, R. E., and Walker, A. 1978. The hominids of East Turkana. *Scientific American 239*: 54–66.

Leakey, R. E., and Walker, A. 1985. Further hominids from the Plio-Pleistocene of Koobi Fora, Kenya. *American Journal of Physical Anthropology 67*: 135–163.

Leakey, R. E., Walker, A., Ward, C. V., and Grausz, H. M. 1989. A partial skeleton of a gracile hominid from the Upper Burgi Member of the Koobi Fora Formation, East Lake Turkana, Kenya. In: G. Giacobini (Ed.), *Hominidae* (Milan: Jaca), pp. 167–173.

MacLarnon, A. 1993. The vertebral canal of KNM-WT 15000 and the evolution of the spinal cord and other canal contents. In: A. Walker and R. E. Leakey (Eds.), *The Nariokotome Homo erectus Skeleton* (Cambridge: Harvard Univ. Press).

Martin, R. D. 1983. Human brain evolution in an ecological context. *52nd James Arthur Lecture on the Human Brain* (New York: American Museum of Natural History), pp. 1–58.

Martyn, J. E. 1967. Pleistocene deposits and new fossil localities in Kenya. *Nature 215*: 476–479.

Noble, W., and Davidson, I. 1991. The evolutionary emergence of modern human behaviour: Language and its archeology. *Man 26*: 223–254.

Pilbeam, D. R., and Gould, S. J. 1974. Size and scaling in human evolution. *Science 186*: 892–901.

Rightmire, P. G. 1990. *The Evolution of Homo erectus* (Cambridge: Cambridge Univ. Press).

Robinson, J. T. 1961. The australopithecines and their bearing on the origin of man and of stone tool-making. *South African Journal of Science 57*: 3–13.

Robinson, J. T. 1972. *Early Hominid Posture and Locomotion* (Chicago: Univ. of Chicago Press).

Rose, M. D. 1984. A hominine hip bone, KNM-ER 3228, from East Lake Turkana, Kenya. *American Journal of Physical Anthropology 63*: 371–378.

Ruff, C. B. 1991. Climate and body shape in hominid evolution. *Journal of Human Evolution 21*: 81–105.

Ruff, C. B., and Walker, A. 1993. The body size and body shape of KNM-WT 15000. In: A. Walker and R. E. Leakey (Eds.), *The Nariokotome Homo erectus Skeleton* (Cambridge: Harvard Univ. Press).

Sarna-Wojcicki, A. M., Meyer, C. E., Roth, P. H., and Brown, F. H. 1985. Ages of tuff beds at East African early hominid sites and sediments in the Gulf of Aden. *Nature 313*: 306–308.

Shipman, P. L. 1986. Scavenging or hunting in early hominids? *American Anthropologist 88*: 27–43.

Skinner, M. 1991. Bee brood consumption: An alternative explanation for hypervitaminosis A in KNM-ER 1808 (*Homo erectus*) from Koobi Fora, Kenya. *Journal of Human Evolution 20*: 493–503.

Smith, B. H. 1993. Physiological age of KNM-WT 15000 and its significance for growth and development of an extinct species. In: A. Walker and R. E. Leakey (Eds.), *The Nariokotome Homo erectus Skeleton* (Cambridge: Harvard Univ. Press).

Stoddart, D. M. 1990. *The Scented Ape: The Biology and Culture of Human Odour* (Cambridge: Cambridge Univ. Press).

Stringer, C. B. 1986. The credibility of *Homo habilis.* In: B. A. Wood, L. Martin, and P. Andrews (Eds.), *Major Topics in Primate and Human Evolution* (Cambridge: Cambridge Univ. Press), pp. 266–294.

Tobias, P. V. 1967. Pleistocene deposits and new fossil localities in Kenya, Part II, the Chemeron temporal. *Nature 215*: 479–480.

Tobias, P. V. 1971. *The Brain in Human Evolution* (New York: Columbia Univ. Press).

Tobias, P. V. 1989a. The gradual appraisal of *Homo habilis.* In: G. Giacobini (Ed.), *Hominidae* (Milan: Jaca), pp. 141–149.

Tobias, P. V. 1989b. The status of *Homo habilis* in 1987 and some outstanding problems. In: G. Giacobini (Ed.), *Hominidae* (Milan: Jaca), pp. 151–154.

Tobias, P. V. 1991. *Olduvai Gorge, Volume 4: The Skulls, Endocasts and Teeth of* Homo habilis (Cambridge: Cambridge Univ. Press).

Toth, N. 1985. Archeological evidence for preferential right-handedness in the lower and middle Pleistocene, and its possible implications. *Journal of Human Evolution 14*: 607–614.

Trevathan, W. R. 1987. *Human Birth, An Evolutionary Perspective* (New York: Aldine de Gruyter).

Trinkaus, E. 1981. Neanderthal limb proportions and cold adaptation. In: C. B. Stringer (Ed.), *Aspects of Human Evolution* (London: Taylor and Francis), pp. 187–224.

Trinkaus, E. 1984. Does KNM-ER 1481A establish *Homo erectus* at 2.0 myr BP? *American Journal of Physical Anthropology 64*: 137–139.

Walker, A. 1976. Remains attributable to *Australopithecus* in the East Rudolf succession. In: Y. Coppens, F. C. Howell, G. L. Isaac, and R. E. F. Leakey (Eds.), *Earliest Man and Environments in the Lake Rudolf Basin* (Chicago: Univ. of Chicago Press), pp. 484–489.

Walker, A. 1991. The origin of the genus *Homo.* In: S. Osawa and T. Honjo (Eds.), *Evolution of Life* (Tokyo: Springer-Verlag), pp. 379–389.

Walker, A., and Leakey, R. E. 1978. The hominids of East Turkana. *Scientific American 239*: 54–66.

Walker, A., Zimmerman, M. R., and Leakey, R. E. F. 1981. A possible case of hypervitaminosis A in *Homo erectus. Nature 296*: 248–250.

Weidenreich, F. 1937. The dentition of *Sinanthropus pekinensis:* A comparative odontology of the hominids. *Palaeontographica sinica 1.*

White, T. D. 1977. New fossil hominids from Laetolil, Tanzania. *American Journal of Physical Anthropology 46*: 197–230.

Wolpoff, M. H. 1971. Competitive exclusion among Lower Pleistocene hominids: The Single Species Hypothesis. *Man 6*: 601–614.

Wood, B. A. 1976. Remains attributable to *Homo* in the East Rudolf succession. In: Y. Coppens, F. C. Howell, G. L. Isaac, and R. E. F. Leakey (Eds.), *Earliest Man and Environments in the Lake Rudolf Basin* (Chicago: Univ. of Chicago Press), pp. 490–506.

Wood, B. A. 1991. *Koobi Fora Research Project, Volume 4, Hominid Cranial Remains* (Oxford: Clarendon Press).

Wood, B. A. 1992. Origin and evolution of the genus *Homo. Nature 355*: 783–790.

CHAPTER 3

ARCHEOLOGICAL INTERPRETATIONS OF EARLY HOMINID BEHAVIOR AND ECOLOGY

■

Richard Potts*

■

INTRODUCTION

FIGURE 3.1

Oldowan chopper from Bed I, Olduvai Gorge, Tanzania.

The modified rock in Figure 3.1 is a stone artifact, an example of the oldest made by early humans. It was manufactured from a lava rock typically available in habitats hominids occupied in Africa between 2.5 and 1.5 million years ago. The sole objective in the **archeology** of early hominids used to be the discovery and analysis of such artifacts. This is no longer the case, at least with regard to the archeology of the earliest known human toolmakers.

The stone objects that early hominid toolmakers found in their habitats were, of course, resources. Just like food and water, they were valuable and useful commodities distributed unevenly in a complex environment. They were things that hominids could use, often to modify foods or other environmental resources. This basic realization that artifacts started out as resources allows archeologists today to move beyond the simple discovery phase to a broader ecological and evolutionary perspective. As resources, the artifacts become relevant to how certain hominids survived and to understanding the conditions faced by these toolmakers. Since natural selection requires an environmental context within which to act, the view that artifacts were extracted from and made in specific, local habitats of early toolmakers enables archeologists to see these simple, chipped stones as possible influences on the evolutionary history of hominids.

Early humans, in fact, were not the first organisms to use stones to modify other things in their environment. Not to confuse the much older age of dinosaurs (about 250 to 65 million years ago) with the much younger period of hominid evolution, nonetheless even certain dinosaurs adopted the use of stones. The **prosauropods** probably were the first vertebrates to swallow and to keep stones in their stomachs to

*Department of Anthropology, National Museum of Natural History, Smithsonian Institution, Washington, DC 20560

FIGURE 3.2

The prosauropods (about 210 to 190 million years old) represent the first radiation of high-browsing dinosaurs. At least some members possessed a gastric mill consisting of pebbles that ground against ingested vegetation in the muscular walls of the digestive tract, thereby facilitating digestion. Illustration by Gregory Paul.

act as a kind of pulp mill useful for grinding up otherwise indigestible plant tissues (Galton, 1985; Figure 3.2). The prosauropods were apparently among the first ground-moving animals of any kind to reach high foliage. By virtue of large body size and anatomical changes enabling an upright stance, some members of this diverse group were able to feed off high vegetation, and the action of their gut stones enabled them to process leaves and other plants that no other vertebrate could. This unexpected example taken from **Mesozoic** reptiles demonstrates that the evolutionary success of a particular group of organisms involved exploitation of a new source of food in a novel manner by using stones from the environment.

In thinking about the evolutionary origin of humans there has always been a significant place reserved for the oldest modified implements. Technology, after all, establishes a range of ecological and economic possibilities for modern human societies. Because technology has been an important element in the history of *Homo sapiens,* even the simplest toolmaking is assumed to have been a key innovation in hominid evolution. According to this standard line of thought, toolmaking had immediate and far-reaching consequences for the survival and success of the genus *Homo.*

Paleoanthropologists have drawn special attention to three important interpretations of the early archeological record that have greatly influenced our ideas about hominid evolution. First, the actual process of making stone tools has been seen as a major innovation of great consequence as soon as certain hominid populations discovered how to do this. Second, the oldest accumulations of stone tools have been interpreted to be the remnants of ancient campsites, places on the landscape where hominids focused their social activities, made tools, shared food, and slept. These

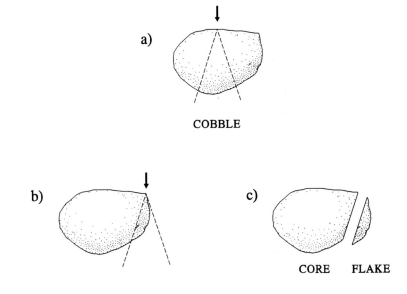

a)

COBBLE

b)

c)

CORE   FLAKE

FIGURE 3.3

Stone percussion and the distribution of force in the manufacture of flakes and cores. (a) When one stone is used to strike another, lines of force radiate out from the point of percussion. (b) If the stone target is hit on a natural edge, only a portion of the cone of force is transmitted through the rock. Because the rock is less massive at this point of percussion, it becomes fairly easy to generate enough force to detach a sliver of stone. (c) The sliver of stone is called a flake, and the struck piece containing the flake scar is called a core. Both will have sharp edges. Flaking and toolmaking in this manner is easiest using rocks fine-grained in texture and possessing the same structural properties in all directions (isotropic).

home bases were equivalent to the campsites of modern foraging people. The existence of such archeological home bases meant that a fundamental step in becoming human had occurred. Finally, the oldest stone tools have been believed to denote the beginning of culture and required some form of linguistic communication to inform and to instruct in their manufacture.

Culture, home bases, and tool manufacture signify basic shifts in human evolution that archeologists have tended to associate with the earliest record of stone artifacts. Recent analyses and discoveries, however, have suggested that we need to reconsider these mainstream interpretations. This chapter examines the changing viewpoints and identifies what has caused the perspective of archeologists to change.

■

THE EARLY ARCHEOLOGICAL RECORD

Artifacts and Sites

Manufacturing the oldest known stone tools primarily entailed finding suitable rocks in the environment, and then making sharp edges on them by breakage. As depicted in Figure 3.3, if two stones are struck together in a certain way, the cone-shaped lines of force transmitted through the rock will detach a sliver of stone, or **flake.** The flake will be sharp almost all the way around its edge. The piece that is struck, the **core**, will have a sharp edge at the place where the flake was removed. When such flakes and cores are found together in places away from ancient river channels or cliffsides where rocks could not have collided or been transported naturally, then we have the crucial information necessary to infer that these chipped stones were the products of hominid hands.

It is quite likely that tools made from rock were not the first tools hominids made or used. Although modern chimpanzees cannot be considered human ancestors, the fact that this closest biological relative of *Homo sapiens* is known to modify twigs, leaves, and other plant objects for use as tools may mean that even the oldest populations ancestral to both humans and chimps possessed the ability and tendency to produce tools under particular circumstances. Wooden implements may have preceded the manufacture of stone tools, yet such plant materials are likely to have perished because their preservation required conditions of fossilization that were rare in the late Miocene and Pliocene environments of sub-Saharan Africa where the first hominids, and the earliest stone toolmakers, apparently evolved.

The oldest accepted bone implements, modified by use and sometimes by stone-on-bone flaking, are less than 2 million years old and are associated with stone tools (Leakey, 1971; Brain, 1985). Some of these bone tools (especially at the South African site of Swartkrans) might have been made by robust australopithecines (Susman, 1988, 1991). This interpretation is consistent with the idea that human and chimp ancestors possessed a general capacity for tool manipulation.

A claim for even older bone toolmakers was made by Raymond Dart (1955, 1957), who proposed that the earliest hominids aggressively applied their bone tool craft to the skulls and flesh of animal and other hominid victims. Dart's **osteodontokeratic culture** (from the Greek roots for bone, tooth, and horn) was based on the discovery of australopithecine fossils (now known to be 3.0 to 3.3 million years in age) in association with the broken **mandibles**, leg bones, and horn cores of antelopes and baboons at Makapansgat, South Africa. The inference that the australopithecines selected and modified certain animal bones for the purpose of murder and carnivorous mayhem is an excellent example of an inference about hominid behavior that has not withstood scientific scrutiny. As shown elegantly by studies of bone assemblages (for example, Brain, 1981), Dart's interpretation did not take into account nonhominid factors that modify and affect the accumulation of animal bones. The particular skeletal parts preserved in the Makapansgat fossil assemblage represent those body parts most durable and likely to survive in any fossil assemblage regardless of the agent of accumulation. Moreover, the damage patterns seen on these bones also suggest the activity of large cats and other carnivores known to have been present in the same environments as the australopithecines. Dart's experiments showed that it was indeed feasible that bones broken in a certain way might be used for battering, breaking, cutting, and potentially violent hominid activities. Feasibility simply based on an ability to reproduce certain behaviors in the present, however, is no longer deemed sufficient grounds for interpretation of hominid behavior. Paleoanthropologists require more substantial, demonstrable connections between the observed details of fossil and archeological remains and the inferences made from them about hominid activity. Studies that showed the great similarities between the Makapansgat bone collections and the effects of nonhominid agents have negated Dart's imaginative possibilities and have forced us to make nonhominid factors an integral part of our interpretation of artifacts and sites.

The oldest known archeological sites, approximately 2.5 million years in age, occur in East Africa, and consist of collections (or assemblages) of stone cores, flakes, utilized rocks, and unmodified pieces of stone raw material that could not have been introduced by geological or other nonhominid processes. These oldest known stone artifacts are called **Oldowan** tools. Hundreds to thousands of fractured stones are found in a cluster, generally in an area 50 to 100 m^2. Such a concentration of artifacts defines an archeological site. This is not to say that hominid artifacts only occur in concentrations; however, such dense clusters of artifacts have long been the primary target of archeological excavation. As we shall see, this traditional focus on dense artifact clusters is now believed to provide only a small window on the way early hominid toolmakers interacted with their habitats.

Over the past 20 years, a great deal of attention has been paid to the **taphonomy** of archeological sites. A taphonomic study evaluates how an assemblage of prehistoric debris formed in the first place, and how the objects were collected, buried, and preserved at the particular excavation site. Taphonomists inquire whether stone artifact sites represent the original places where hominids dropped stone tools, or were places disturbed by other forces such as rapid streams that could move stone tools originally left by hominids upstream.

With regard to the earliest archeological sites, the answer to this question is that both disturbed and undisturbed contexts of preservation (taphonomic contexts) have been found. Some sites, such as the Kada Gona 2-3-4 site at Hadar, Ethiopia, preserve cores and flakes characteristic of hominid stone-on-stone percussion in association with cobbles, sands, and sedimentary evidence of a stream channel (Harris, 1983). Such sites are evidently the disturbed remnants of tool assemblages made by hominids somewhere in the vicinity. Other sites, however, can be demonstrated to be relatively undisturbed places where hominids visited and made and discarded stone tools. Many of the sites in Bed I at Olduvai Gorge are examples of such delimited areas where hominid toolmakers were present and geologic forces primarily acted to preserve, rather than to disrupt and delete, the traces of hominid activity (Leakey, 1971; Potts, 1988). As we shall see, this does not negate the possibility that agents other than hominids contributed significantly to site formation.

Although stone artifacts provide the basis for identifying Oldowan archeological sites, associated animal bones are also often very abundant. These fossils usually represent parts of animals transported away from the death sites. In many cases, hominids obtained these bones and brought them to the sites where the artifacts and bones of many other animals were accumulated. Although they have long been presumed to be the food remains of hominid toolmakers, only over the past dozen years have techniques been developed to distinguish marks on these fossil bones made by stone tools from damage caused by the teeth of carnivores, sedimentary particles, excavation, and other agents (Figure 3.4).

The animal bone assemblages excavated from Olduvai and Koobi Fora tend to consist mainly of meat-rich bones, and many but not all of the cut marks were caused by slicing action in areas of muscle attachment. Besides these definite signs of meat acquisition, there is evidence that hominids also broke the long-shafted leg bones of animals to acquire marrow. Stone tool percussion marks have been detected on these leg bones, and middle or shaft portions tend to be fractured all the way down to their ends, as humans do when using stone tools to obtain marrow (Potts, 1988; Bunn, 1989; Blumenschine and Selvaggio, 1988). Cut marks and traces of stone tool breakage of long bones are exactly the kinds of evidence needed to implicate hominids securely as collectors and users of the animal bones. As explained below, however, early archeological sites often have a complicated taphonomic history, in some cases with evidence that large carnivores were also attracted to the exact same places on the landscape.

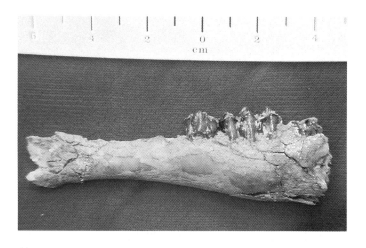

(a)

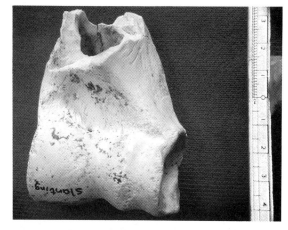

(b)

(c)

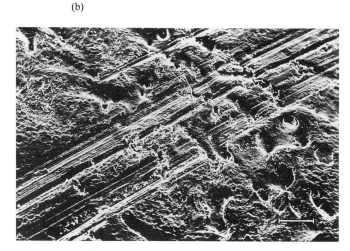

(d)

FIGURE 3.4

Animal bones found associated with Oldowan sites from Olduvai Gorge, Tanzania. (a) Bovid (antelope) jaw; (b) Bovid (antelope) distal humerus; (c) Bone fragments. Most of the bones are broken in some way, and many fragments are often found. The bone in (b) shows grooves that have been determined to be cut marks by microscopic examination. (d) A scanning electron micrograph of the fine microscopic striations that distinguish stone tool cut marks from the smooth contour of grooves made on bones by carnivore teeth. Micrograph bar scale equals 0.2 mm.

SETTING HOMINIDS IN AN ECOLOGICAL CONTEXT

What Do We Want to Know?

What exactly do we want to learn about hominids? What information would we need in order to ascertain the most important aspects of hominid evolution? With regard to any organism, survival and differential reproduction are based on the behavior of individuals, the behavior strategies adopted by social groups, and the pattern of interactions within ecological communities. Hypothetically, a vast array of behavioral and ecological parameters contributed to the evolutionary history of a species and its populations. These parameters can be grouped into several related categories.

1. *Feeding.* Because food provides energy and nutrition, the parameters classified in this category are fundamental to survival and reproduction. The parameters include food choice, food quality, the time and energy costs of food search, and the means of search (for example, hunting versus scavenging by meat eaters). For hominids in particular, resources other than food—such as stone tools, digging sticks, and carrying bags—may be involved in feeding. Acquiring and producing these items takes time and energy and thus contribute to the total advantages and costs involved in feeding. Further, competition either within or between species may affect an individual's ability to acquire particular foods (Pianka, 1983; Roughgarden et al., 1989; Krebs and Davies, 1991). Competition for resources may be expressed indirectly by exploiting a resource before the potential competitor arrives on the scene (exploitation), or directly by confrontation (interference). Finally, diet and foraging are affected by the availability and distribution of foods over time and space.

2. *Predation.* Susceptibility to predation is influenced by several parameters, including the duration and location of activities such as feeding that are performed within a predator's habitat, and the behaviors such as avoidance, protection or aggression by which organisms deal with potential predators.

3. *Mating.* Like food, and rocks for tools, mates may be viewed as resources. Search time and intraspecies interactions are parameters that affect mate choice, reproductive rate, and relative reproductive success. The size and organization of the social group are influenced by the mating system and the degree to which parents invest in young, all of which appear to be affected strongly by food distribution (Rubenstein and Wrangham, 1986).

4. *Demography.* Population size, density, age and sex composition, shifts in group size, group movements, and home range size are included in this category. These parameters influence and, in turn, are affected by feeding, mating, and predator avoidance. Population size, structure, and movements partially determine or influence the effects of natural selection, random processes of genetic change, and gene flow.

5. *Life history traits.* Certain characteristics of organisms do not fall under the heading of behavior or group dynamics, but reflect aspects of growth and maturation over the life cycle. Evolutionary biologists have paid increasing attention to such life history characteristics. These include gestation length, growth rate, body size, longevity, and metabolic constraints—traits that obviously assert strong influence on behavior and that evolve in relation to ecological conditions and social dynamics (see Lovejoy's discussion in Chapter 1).

What Can We Expect to Learn?

Having defined a set of important evolutionary parameters, the second step in this analysis concerns what information one can expect to obtain from archeological remains. Usually, few of the above ecological parameters can be given due treatment even by field biologists who have the ecology and behavior of living species available to examine directly. Because archeological materials consist of stone artifacts and associated animal remains, we may phrase the question as follows: What information of evolutionary significance is potentially obtainable from bone and lithic materials excavated from known sedimentary contexts?

First, geological analyses can yield reconstructions of sedimentary environments, the paleogeography of ancient habitats, the overall vegetation structure (for instance, relative percentage of **C₄ grasses** versus **C₃ woody plants**), and climates. Analysis of fossil pollen when available and stable **isotopes** of soil **carbonates** can provide more direct information on plants and vegetation structure of old landscapes. Estimates of rainfall may be derived from stable isotope studies of carbonates and from **palynology** (Cerling and Hay, 1986; Bonnefille and Riollet, 1980). Data comparisons over time and space help to detect variation in paleoecological parameters that were likely to have influenced hominid survival.

Second, faunal remains have long been used in paleoenvironmental reconstruction. By identifying taxa from bones or shells, individual species with precise environmental tolerances, or whole communities comprising many vertebrate and invertebrate species, inferences may be drawn concerning ancient habitats. Certain taxa may be sensitive indicators of ecological conditions, such as climate and vegetation. Fossil fauna and geology have consistently documented a mosaic of habitats associated with early hominid evolution. Grassland, woodland, closed bush, open floodplains, and lake margins (including swamps and marshes) penetrated by channel systems and riverine habitats (such as gallery forests) have been described as an environmental complex that typified the late **Tertiary** and **Quaternary** of East Africa (Van Couvering, 1980; Andrews, 1989; Bonnefille, 1985; Hay, 1976; Laporte and Zihlman, 1983). Although the inferred diversity of habitats does suggest the overall milieu in which hominids lived, it provides no clues as to specific resources, habitats, or selective pressures relevant to hominids. Ideally, concepts of resource distribution, availability, and temporal variation should supplant simple habitat categories such as "savanna" or "woodland" (Winterhalder, 1980). Only recently, however, have archeologists begun to sample wide areas of paleolandscapes in order to assess the spatial relationship of hominid toolmakers to habitats (Potts, 1989b, 1991; Blumenschine and Masao, 1991). Because climates and resources exhibit seasonal and annual variations in African savannas today, time resolution on the order of months or years would provide useful information about environmental heterogeneity and possible behavioral responses. The resolution afforded by early archeological data, however, appears to be poorer than this; the only attempt so far to determine the quality of time resolution suggests that a minimum of several years may have been involved in the accumulation of archeological assemblages in Bed I of Olduvai, and there is no indication that hundreds of years might not be involved in the formation of some of these sites (Potts, 1986, 1988).

Nonetheless, where archeological bone assemblages can be attributed to hominid foraging, the faunal remains provide an idea about animal resources used by the toolmakers. Ecologically sensitive species in such assemblages would indicate specific habitats used by hominids and possible frequencies at which they were used, either as a result of their relative availability or preferred exploitation by hominids. The bones may potentially indicate the ways animal tissues were obtained, for example, by hunting or scavenging. Although methods have been developed to determine which parts of animal carcasses tend to be available for scavenging (Blumenschine, 1986, 1987), at this point it is still difficult to distinguish hunting from scavenging on the basis of bone assemblages at archeological sites, especially if the scavenging involved first or early access to an animal freshly dead. Unless hunting can be distinguished from scavenging, the degree to which animals in archeological faunas represent predator-prey interactions cannot be evaluated.

It is known that certain hominid toolmakers did handle animal tissues; competitive overlap with carnivores was thus likely. East African habitats are notable for the great diversity of hunting and scavenging species that occur **sympatrically**, and may have competed directly for the same carcass. By virtue of living in savanna-mosaic areas, predation on hominids by large carnivores was a clear possibility. Yet rarely has the nature and intensity of ecological overlap between hominids and carnivores been evaluated (Potts, 1988).

Apart from the environmental evidence compiled by geologists and faunal experts, a further significant element of hominid behavior may be judged from the aggregated nature of archeological debris. Hominids carried animal tissues and stone artifacts to defined places on the landscape. This implies that specific ecological or social behavioral reasons underlie these activities and that investigation of this phenomenon may provide important clues pertaining to hominid evolution.

Tool production by flaking stone entailed acquiring various stone raw materials. The locations of rock sources on paleolandscapes, especially large outcrops, may be identified. This information furnishes a way to estimate the distances hominids traveled for this specific resource (minimum home ranges), the possible habitats visited during trips to acquire stones, and the possible energetic costs of transporting raw materials or tools made from them.

Finally, the stone tools themselves should provide helpful clues to answer some of our questions about hominid evolution. Ironically, even though stone tool shape and manufacturing methods have been intensively studied, they have been examined only infrequently from an ecological perspective. What can the tools tell us of resource utilization, the use of specific habitats, the ecological advantages of specific artifact attributes and technologies? Theoretically, the uses of artifacts are possible to discern, such as by microwear analysis of tool edges (Keeley, 1980). Inferences about artifact usage would provide some idea about the environmental resources used by hominids. At least indirect inferences about hominid diet, for example, could derive from identifying the kinds of plants or animal tissues responsible for wear on the edges of stone tools. In general, though, the uses of **Paleolithic** artifacts by their hominid makers are hardly known. The probable uses of less than a dozen artifacts from the early **Pleistocene** of Africa have been assessed (Keeley and Toth, 1981). Until extensive analysis of artifact edge function is accomplished, clear answers are not foreseeable concerning how specific tool types, artifact attributes, or technological changes enhanced hominid use of their environments.

It should be clear that excavated archeological remains potentially contribute information on a small but valuable subset of parameters relevant to early hominids and their evolution. Archeological data are most useful for assessing environmental setting, hominid diet, foraging, other aspects of resource use, and the implications of resource accumulation in delimited areas. No hypotheses, however, about mating, most demographic characteristics, and most life history characteristics are likely to be tested using Plio-Pleistocene archeological data. Perhaps **paleontological** data on the fossil hominids themselves, or comparative studies of living primates, can begin to address these problems (Lovejoy, Chapter 1; Walker, Chapter 2). Moreover, some important questions about paleoenvironment and resource use may lie outside the traditional range of early archeological data. For example, overlap in resource use between hominids and other animals would be impossible to infer using evidence pertaining solely to hominid activities.

Despite these baseline limitations, early archeological remains have fueled numerous distinctive and significant ideas about early hominid behavior and evolution. We will summarize how a number of these far-reaching ideas have become subject to reevaluation.

FIGURE 3.5

FIGURE 3.5

Map of Oldowan stone artifact sites.

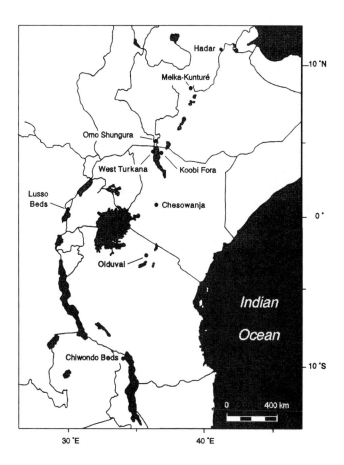

THE SIGNIFICANCE OF MAKING TOOLS AND MOVING RESOURCES

The Handy Man Hypothesis

The oldest stone tools discovered before the 1930s were teardrop-shaped objects known as **handaxes.** The manufacture of stone handaxes, which we will explore later, typifies the stone technology or industry known as the **Acheulian.** Largely through Louis Leakey's energetic vision of Africa as the source of the earliest hominids, stone tools cruder and older than Acheulian handaxes were discovered and documented from the 1930s to 1950s at Olduvai Gorge, Kanjera, and Kanam (Leakey, 1935). These artifacts were termed *Oldowan* after the original spelling of Olduvai Gorge (Oldoway). Assemblages of early Oldowan artifacts have now been uncovered from many sites in East Africa in particular (Figure 3.5, Table 3.1). The fact that archeologists could distinguish diverse shapes among the flaked Oldowan stones suggested that even the earliest hominid toolmakers imposed certain tool designs on the lumps of rock they modified, and that these designs must therefore have been reproduced by cultural rules.

The naming of *Homo habilis* ("handy man") by Leakey, Tobias, and Napier (1964) represented a formal proposal that linked stone toolmaking to a new definition of the genus *Homo*. In fact, for the first time in the history of paleontology, a behavior — the manufacture of stone tools — was strongly considered in the naming and formal

▬

TABLE 3.1

Excavated Oldowan sites, 1.5 to 2.5 million years in age, showing locality, age, excavations, artifacts, and fauna. Secure causal associations between artifacts and fauna based on tool cut marks have been established so far only in Bed I Olduvai and Koobi Fora. For more information see Potts, 1991.

Locality	Age (in millions of years)	Excavations	Artifacts[1]	Associated Fauna[1]
Hadar	2.5–2.7	Kada Gona 2-3-4	p	–
	2.4	Kada Gona West	p	–
Omo Shungura Member E[2]	2.4–2.5	Omo 84	+	–
Omo Shungura Member F	2.3–2.4	FtJi 1	+	+
		FtJi 2	+	–
		FtJi 5	p	p
		Omo 57	p	p
		Omo 123	+ + +	–
West Turkana	2.3–2.4	Lokalelei	p	p
	~1.8	Kokiselei	+	+
	~1.7	Naiyena Engol	+	p
Lusso Beds	2.0–2.3	Senga 5A	+	+ + + +
Chiwondo Beds	>1.6	Mwimbi	p	–
Koobi Fora, Lower Member	~1.9	3 excavated sites	+	– to + +
Koobi Fora, Upper Member	1.5–1.6	18 excavated sites	+ to a	– to + + + +
Olduvai Beds I and II	1.9–1.6	14 excavated sites/levels	p to + + + +	p to a
Melka-Kunturé	~1.5	Gomboré IB	a	+ + + (?)
Chesowanja, Chemoigut Fm.	~1.5	GnJi 1	+ +	+ +
		GnJi 2	+	

1. Abundance of artifacts and faunal remains: –(0–10 specimens); p (present, 11–100 specimens); + (101–500 specimens); + + (501–1000 specimens); + + + (1001–2000 specimens); + + + + (2001–5000 specimens); a (abundant, >5000 specimens).

2. Unconfirmed stratigraphic position and age.

definition of a fossil species. In defining this earliest representative of *Homo*, Leakey's interest in stone tools, Tobias's interest in brain evolution, and Napier's interest in the evolution of the hand became integrated. Their work together reflected theoretical assumptions that have had a pervasive influence on the thinking of paleoanthropologists. They posited strong connections among several critical elements in human evolution. These included tool manufacture, development of manipulative skills, cerebral organization of fine hand coordination, and the development of mental skills and social arrangements needed to produce artifacts and to reproduce over many generations the craft of toolmaking. All of these elements were now linked inextricably to the origin and evolution of the genus *Homo*.

Articles by the influential American paleoanthropologist Sherwood Washburn (for example, 1960) set the stage for tying together these elements of human behavior. Washburn argued that all of these were linked in a tight feedback relationship, whereby change in one amplified the rate of evolution in the others. The making of tools was a prime mover in this whole system, having ramifications for the evolution of the brain, intelligence, language, and culture, and the evolution of the hand and ability to use it in fine skills. Thus, the oldest evidence of toolmaking in the archeological record held enormous importance in early hominid research. The title of Kenneth Oakley's popular book *Man the Toolmaker* (Oakley, 1961; first published in 1949) elegantly reflected the theoretical lens through which paleoanthropologists saw and interpreted the earliest stone tools.

New Analyses of Oldowan Artifacts

The simplest kinds of tools we might envision would include (1) natural objects modified solely by using them, and (2) objects prepared for use by a single basic procedure. This is exactly what we see in the tool-using activities of chimpanzees (*Pan troglodytes*). Certain populations of chimps in West Africa are known to use stones as hammers and anvils for cracking nuts. These stones are altered simply by their recurrent use in this activity. Chimpanzees at Gombe Forest, Tanzania, and elsewhere are also known to modify twigs by stripping leaves and smaller branches, which are then inserted into termite mounds to catch the ants for consumption. Young individuals observe older chimps doing this activity; however, a considerable amount of trial and error may be involved in replicating this behavior. The result is the ongoing reproduction of a behavior distinctive to chimpanzee groups in certain areas, but not seen in others (Goodall, 1986; McGrew et al., 1979; Boesch and Boesch, 1984).

The oldest assemblages of stone tools associated exclusively with bipedal hominids, not with chimpanzees or other apes, also exhibit these two kinds of modification. Sharp-edged flakes and cores, for example, are produced by the simple procedure of striking one stone on the edge of another. **Hammerstones** are an example of the other form of simple implement, modified solely by battering. All of the objects made in the Oldowan technology reflect a gradation between these two basic activities—the repetitive making of sharp edges and the battering of stone surfaces (Toth and Schick, 1986; Potts, 1991).

The flaked pieces in Oldowan artifact assemblages can be divided into specific categories with names such as chopper, scraper, and discoid (Leakey, 1971). These names do not specify intended purposes or actual functions, but rather simply the patterns of flake scars on these stones. These classes of tools and cores were once assumed to reflect actual target designs of hominid toolmaking, tool types imposed by the cultural rules passed on by hominids.

Research over the past decade or so, however, has negated this idea. Rather than being discrete, exclusive artifact types, the shapes of Oldowan tools instead appear to be continuous, overlapping expressions of the basic procedure of creating sharp edges by stone percussion. Figure 3.6 illustrates the continuities among the most common flaked pieces. Oldowan choppers, discoids, heavy-duty scrapers, polyhedrons, and subspheroids evidently were various incidental stopping points in the process of removing flakes from cores (with the addition of battering in the case of subspheroids). This has now been shown by detailed measurement of these flaked pieces. Their mutual overlap in size and edge length is extensive (Figures 3.7 and 3.8). The average number of flake scars also varies; it goes up for more modified pieces, but there is a gradual continuum between one core and another in this characteristic (Table 3.2). No matter which attribute is selected, the pieces form a continuum. The flaked stones of the Oldowan thus cannot be demonstrated to constitute

FIGURE 3.6

Continuities among Oldowan artifact types. Removal of a few flakes transformed a cobble into a chopper and then, if more flakes were struck, into a discoid. Similarly, a split cobble, heavy-duty scraper, and polyhedron form a continuum. The basic Oldowan core types result from applying the single procedure of percussion flaking to stones that were diverse in their original shape.

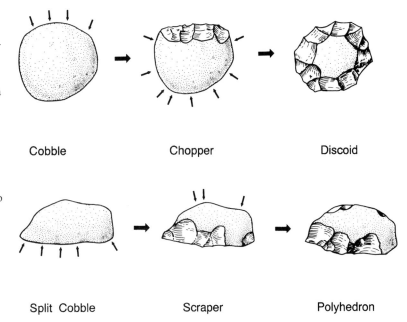

Cobble   Chopper   Discoid

Split Cobble   Scraper   Polyhedron

discrete target designs but can be shown to represent simple by-products of the repetitive act of producing sharp flakes (Potts, 1988, 1991; Toth, 1985, 1987; Toth and Schick, 1986).

Since the discovery and naming of *Homo habilis*, a number of findings has led us to reassess whether Oldowan toolmaking was truly the key adaptive innovation and distinction between *Homo* and the australopithecine hominids it has generally been construed to be. First, since the mid-1960s, observations of various chimpanzee groups in their natural habitats have documented the use of stones, twigs, and other materials as tools by these apes (for instance, Goodall, 1986; Boesch and Boesch, 1984; McGrew, 1989). Second, a series of articles by Randall Susman (1988, 1991) has recently suggested that hominids, including the australopithecines, generally had the anatomical capacity to make and to manipulate tools. If this is so, then toolmaking per se cannot have constituted the main "adaptive wedge" driving the evolution

FIGURE 3.7

Diagram of weight of five "types" of flaked pieces in Oldowan assemblages from Bed I, Olduvai Gorge.

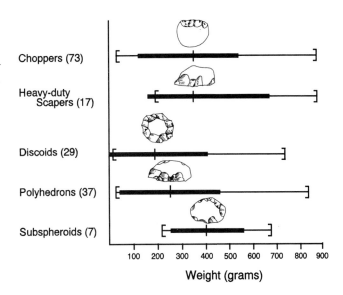

FIGURE 3.8

Diagram illustrating overlap in the length of the flaked edge (striking platform) of four "types" of flaked pieces in Oldowan assemblages from Bed I, Olduvai Gorge.

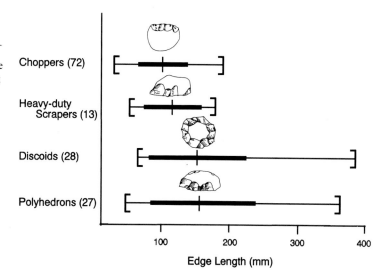

of hands, brains, and behavior in early *Homo*, separating these toolmakers from other bipedal and ape relatives. Finally, as we have seen, the oldest known stone technology is now perceived to have been exceedingly simple, and not made up of culture-bound designs for making tools held in the minds of toolmakers. All of these points have created doubts about whether the actual process of making stone tools was the primary, significant element in the survival and evolution of early hominid toolmakers.

The Resource Transport Hypothesis

If making tools, in itself, was not such a monumental development, what adaptive and evolutionary consequences resulted from early stone tools? Why did some populations of hominids become engaged in making lithic artifacts? What were the benefits of this behavior?

TABLE 3.2

Number of flake scars found on four major types of flaked pieces in Oldowan assemblages from Bed I Olduvai Gorge.

Type	N	Total Estimated Scars	Mean Flake Scars	Standard Deviation
Choppers	66	356	5.4	1.6
Heavy-duty scrapers	12	65	6.4	3.1
Discoids	19	167	8.8	2.6
Polyhedrons	9	64	7.1	1.7

FIGURE 3.9

Viewed from the famed FLK "Zinj" excavation at Olduvai Gorge, the hill in the distance is the Naibor Soit inselberg, an outcrop where hominids probably obtained quartzite for making tools. This outcrop is about 2 km north of the FLK site, where the vast majority of stone pieces are composed of quartzite. These quartzite pieces tend to be small. They make up nearly 28% of the total weight of stone material in the artifact assemblage from this site (>72 kg). Various types of volcanic lava represent 72% of the assemblage by weight, and a few, small artifacts of gneiss and feldspar comprise only 0.1% of the total weight of stones (Potts, 1988). The probable source of gneiss was approximately 8 km from FLK.

Regional geologic studies in the vicinity of Oldowan sites have located some of the stone outcrops used by the toolmakers. The distances between the sites and raw material sources suggest reasons why hominids would not have become involved in making stone tools. At Olduvai Gorge, for instance, the distance separating Oldowan activity sites from the stone sources varied from 2 to 12 km (Figure 3.9). Stones of different kinds of raw materials were brought to sites from various directions. Both the unmodified raw materials and flaked pieces were irregular and bulky to carry, and the total accumulation at any given site was considerable (12.0–93.9 kg for 4 sites in Bed I, Olduvai [Potts, 1988]). These costs of transporting rock must be kept in mind in evaluating why stones became part of the persistent survival strategies of certain hominids.

An important benefit of Oldowan toolmaking would seem to reside in the uses of the tools. As we mentioned earlier, cut marks and signs of tool breakage of animal bones indicate that Oldowan artifacts (at least after 1.9 million years ago) were used to slice and bash bones to obtain meat and marrow. The only direct study of Oldowan artifact edges carried out to date, however, has isolated distinctive microscopic polishes on only nine artifacts (Keeley and Toth, 1981). All of these artifacts are flakes from Oldowan sites at Koobi Fora. The polishes convincingly match those used in cutting meat, wood, and soft plant material. Artifacts in that study are not the oldest known—only about 1.6 million years in age—but they do show that the sharp-edged slivers derived by stone percussion were very useful to Oldowan toolmakers. Systematic studies of this kind have not yet been carried out in older Oldo-wan assemblages, largely because the predominant composition of these earliest artifacts is lava and techniques have yet to be developed to see microscopic polish or damage on lava stone edges adequately.

Just from the evidence of bone modification, however, it is apparent that hominids were able to gain access to the carcasses of large mammals and possibly other new food resources by virtue of making and using sharp stone implements. When making tools out of stone was first tried, we might imagine that hominids took these tools directly to the food sources, and the assistance provided by the tools in obtaining food may have proved a considerable benefit. This manner of foraging, nevertheless, would not have greatly varied from the tool-assisted foraging behaviors of chimpanzees, as exemplified by the transport of a hammerstone to a tree where nuts are available to crack open (Boesch and Boesch, 1984). This means that the act of making or using simple stone tools did not necessarily dictate a more complicated movement of resources, such as the dual movement of distant stones and animal bones to common locations.

Yet by 2 million years ago, this complicated movement of resources is exactly what was practiced by Oldowan toolmakers. An increasing body of evidence shows that Oldowan hominids moved stone raw materials, flakes, and modified pieces in and out of concentration areas (Toth, 1987; Schick, 1987; Potts, 1988, 1991). Besides this movement of stones, animal bones were also repeatedly and extensively transported by the toolmakers. Certain hominids moved both stones and portions of large animal carcasses around the environment, and they tended to move them to the exact same places on the landscape. This transport aspect of the Oldowan highlights other possible advantages gained by the toolmakers that helped offset the potentially large costs of transporting stones.

A new idea may be proposed called the *resource transport hypothesis* (Potts, 1991). According to this idea, a shift in how hominids used resources over the landscape—that is, the spatial use of resources—ultimately comprised the most significant innovation of the Oldowan. Certain groups of hominids began to exhibit behaviors beyond the most basic toolmaking skills. These behaviors reflected a more complex use of the landscape than seen in any contemporaneous hominid or nonhuman primate. Stones were carried repeatedly from their sources, across considerable distances, to particular spots on the landscape. Originally, any given spot may have been a location of an animal carcass or another convenient place where stone tools were used. These sites became nodes out of which the products of stone flaking were carried away and more raw materials were introduced. Animal carcasses found in ever-changing locations were also brought (over a period of at least several years, and perhaps much longer) to be processed at these same places. Over time many stone artifacts and animal bones ultimately became aggregated. By 2 million years ago and possibly before, Oldowan toolmakers evidently had solved the problem of getting two resources together, food that was inconsistent in its location and the stone needed to process the food. Once this behavior was engaged, essentially any resource that could be picked up and moved was made available for processing by stone tools.

This innovation in foraging, made possible but not required by Oldowan toolmaking, is likely to have had several major consequences. The mental mapping of stone and food resources on to one another would have become much more complex than is required by the foraging behaviors known in other higher primates. The combined moving of both animal parts and stone artifacts meant that the diversity and spatial structure of resources hominids encountered and remembered must have been amplified. We may also surmise that there were other consequences pertaining to the movement of hominids around their environments. It could be that this new way of using the landscape was an important precursor facilitating the shifts in locomotor anatomy and body proportions that coincided with the origin of *Homo erectus*. Carrying stones, searching out mobile foods (scavenging and hunting for animals), transporting them to processing sites, and the attention of predatory carnivores that this would have stimulated all may have been a significant antecedent to the evolution of a large-bodied, diurnal, sweaty, long-distance walking hominid of the kind we now strongly suspect early *Homo erectus* represented (Walker, Chapter 2).

Home Bases

Over the past 15 years, a great deal has been written about the home base interpretation of early archeological sites. Archeologists traditionally have looked to social factors to explain why hominids introduced animal bones repeatedly to specific places on the landscape. The early Oldowan sites have been interpreted to be home bases, similar in function to the camping sites of modern human foragers who gather and hunt food. Accordingly, these sites have been viewed as safe, social refuges where hominids developed most of the framework of social interactions, food sharing, division of labor, language, prolonged maturation, and cultural learning typical of modern humans (for instance, Isaac, 1978).

The standard assumption that the oldest collections of stone artifacts and broken animal bones represent home bases has been reassessed from two perspectives recently. First, detailed taphonomic studies have shown that the famed sites from Bed I Olduvai, once believed to be the best cases of early hominid home bases, were places where both hominids and large carnivores were active. Insofar as hominids were the primary collectors of the animal bones, carnivores were also attracted to meaty and marrow-rich remains of animals left behind and left unprocessed by hominid toolmakers. Just as stone tool cut marks have been distinguished microscopically, so too have patterns of bone modification characteristic of hyenas and other carnivores been discerned on the fossil bones from these same Olduvai sites. Because of this exact spatial overlap between hominids and potential predators and their mutual attraction to the same transported parts of animal carcasses, it seems unlikely that these early sites at Olduvai were the safe, social home bases that have typically been envisioned. Given that both hominids and carnivores fed off the food remains at these particular sites, the occurrence of social food sharing as the main purpose behind hominid transport of animal bones to these sites also cannot be demonstrated. Archeological shelters, hearths, or other visible benchmarks of modern forager campsites also are not known with certainty before the Middle Pleistocene (Potts, 1988).

A second reason why the home base interpretation has been reconsidered derives from the recent proposal of alternative models of hominid land use. According to the routed foraging model (Binford, 1984), hominids would have regularly stopped at the same watering holes, sleeping places, and other spots where fixed resources existed. As a result of recurrent attraction of toolmakers to these fixed points on the landscape, artifacts and animal bones became aggregated over time.

An alternative model is the stone cache interpretation, which derives from study of the Oldowan sites of Bed I at Olduvai (Potts, 1988). According to this model, tools or raw materials were carried into the foraging range and were deposited by hominids originally where a carcass was processed or near some convenient resource. These places were revisited by hominids because stones could be obtained there that were needed to process foods found nearby. Because these various places in the foraging range were remembered as locations where stone tools were left, more stones and animal parts were brought in and accrued over time. The presence of transported stone materials thus attracted hominids repeatedly to these sites. A stone cache refers not to a planned strategy of stone collection followed by foraging, but to initial, incidental deposition of stones in multiple parts of the foraging range which, as a result of repeated transport and removal of stone artifacts and animal parts, develop into aggregates of modified and unmodified artifacts and bones. The stone cache interpretation of sites from Olduvai Gorge was originally inspired by computer simulations of the effort (energy) entailed in transporting stone and bone to the same spots. The simulations illustrate that transport of animal bones to multiple sites where stone material is available represents an efficient method of bringing these two resources together. The stone cache model thus is an example of what might be termed multiple place foraging.

The home base model still lives in modified form, now called the central place foraging model (Isaac, 1983, 1984). It draws its inspiration by analogy to hunter-gatherer campsites and continues to denote the occurrence of the same basic combination of human social and foraging behaviors once implied by the home base hypothesis (division of labor in acquiring meat and plants, food sharing, brain enlargement with enhanced communication, male investment, and pair bonding). Rather than fixed environmental resources or the nearby proximity of stones needed to process foods, the central place foraging idea stresses campsite-type social behavior as the main magnet attracting hominids to sites.

The existence of multiple behavioral models in recent years has freed archeological thinking from strict reliance on the home base interpretation. It means that the origin and development of home base behaviors may actually be detected archeologically rather than simply imputed to the oldest occurrences of clustered artifacts and associated animal bones. As noted previously, we know that at least certain Oldowan sites signify a radical innovation—namely, that hominid stone toolmakers by 2 million years ago had resolved the basic problem of getting dispersed animal food sources and processing tools together at the same time. This alone would have been an exceedingly important antecedent to the later development of home bases and home base social behaviors.

There are two ways to envision the development of home bases. In the first scenario, a social focus was part of the original grouping behavior of hominids. The development of home bases entailed the grafting of transport behaviors on to the spatial pattern of social movement in these hominid groups. Because of the attraction of predators, animal parts (and stones to process them) were not brought at first by hominids to a central place social focus. Ultimately, though, food and stone tools were carried repeatedly to the pre-existing node where the social group gathered, and this resulted in the formation of home bases.

The second scenario derives from the fact that shifts in the location of resources influence the grouping behavior of males, females, and young in any social community. Resource transport, that is, solving the problem of getting stone and food resources together, had the capacity to cause such a change in grouping behavior. The consistent, daily attraction of individuals to the place of resource transport is unlikely to have happened as long as the potential for carnivore attraction to these same sites was high. Once this changed by minimizing attractive leftovers, controlling fire, or other means of site defense, however, the preexisting nodes where resources were transported would have attracted members of the social community on a daily basis, resulting in the formation of home bases and associated behaviors of daily movement seen in modern human foragers.

■

NEED FOR A NEW CONCEPT OF CULTURE

Repetition in the Behavior of Early Hominid Toolmakers

The concept of culture may well be anthropology's most enduring and pivotal contribution to the social sciences. The making of old stone tools has long been taken to reflect the beginning of culture and fundamental shifts in behavior. Such an interpretation was based originally on the assumption that the tools were target designs, the intended products of hominid minds guided by social learning, and also that making Oldowan tools required learning through verbal language. The use of the term culture, or cultures, to refer to early Paleolithic assemblages of stone tools echoes this assumption. It implies that the fundamental processes of human cultural behavior are mirrored by the manufacture of lithic artifacts, and that toolmaking represents a recognizable point of origin of this entity called culture.

As we have already seen, this pervasive view has been contradicted by new analyses of old stone tools. Oldowan artifacts cannot be considered separate designs. They are fluid expressions of removing flakes and altering stones by using them. A second point of contradiction can be seen in the homogeneity of artifacts. Over the 200,000 year record of the Oldowan at Olduvai Gorge, essentially no change can be seen in the products of toolmaking behavior, in site production, or in any other behavior reflected in the material evidence from archeological sites (Leakey, 1971; Potts, 1988). Hominid toolmakers kept replicating the same behaviors over this long period.

This point can be demonstrated further by looking at an even later expression of hominid toolmaking, still the early Paleolithic but after a new form of artifact had been conspicuously added to the repertoire of hominid toolmaking. This artifact is the **biface**, the most abundant example of which is called the *handaxe*. Handaxes first occur in the archeological record about 1.5 million years ago, more than a million years after the oldest known instances of stone toolmaking. Handaxes then became the dominant component of stone toolmaking over the next one million years or so, spanning Africa, much of Asia and Europe, as hominids expanded their geographic range.

The site best known for handaxes in Africa is Olorgesailie, located in the rift valley of southern Kenya (Isaac, 1977; Potts, 1989b). Figure 3.10 displays some of the variation in handaxes known from this site. Handaxes of the Acheulian industry were generally made from large flakes (longer than 10 cm) struck from a boulder. Because the manufacture of large flakes is something not seen in the Oldowan, this step appears to have been an important breakthrough leading to the regular manufacture of Acheulian handaxes. After such large flakes were produced, handaxes were fashioned by chipping all around the circumference. Ovate symmetrical pieces, sometimes pointed at one end, were the result.

For the first time in stone technology, hominids apparently were producing an artifact of a particular target design (Isaac, 1972, 1977). And they kept on producing it. Figure 3.10a–c shows three handaxes from Olorgesailie fashioned over a period from about one million to 600,000 years, based on single crystal **argon dating** (Deino and Potts, 1990). They were all made out of locally available lava materials. They were all made from large flakes. They show flake scars around the periphery and across both top and bottom surfaces. They are all ovate, somewhat symmetrical, and are pointed at one end. Other examples from Olorgesailie are somewhat less pointed. To illustrate further the variation in handaxes, Figure 3.10d is a handaxe from Eu-

(a)   (b)   (c)   (d)

FIGURE 3.10

Four examples of Acheulian handaxes represent much of the range of variation in this tool category during most of the Pleistocene. (a–c) are from Olorgesailie, Kenya, and (d) is from Europe.

rope, a little later in time and made of chert, and it too is flaked all around the perimeter, is pointed at one end, and forms that teardrop shape typical of handaxes. The act of making handaxes was amazingly repetitive. Hundreds of thousands of other examples could be illustrated, derived from a period over 1 million years long, convincing evidence of the utter monotony of early hominid toolmaking during most of the Pleistocene.

Several years ago during a professional conference in honor of J. Desmond Clark, one of the deans of early hominid archeology, Professor Clark was introducing the speakers and papers being presented by his colleagues. The audience had just heard two papers back to back that concluded on the basis of brain endocasts that *Homo habilis* and *Homo erectus* possessed the ability to speak, and that they used language as a way to communicate how to make stone tools. Professor Clark stood up and commented, "If those early hominids were speaking to one another about tools, they must have been saying the same thing over and over again for a very long time."

That is indeed the point. Language as a means of transmitting cultural information involves a rich, inevitable creation of new utterances by using symbols that can be put together in extremely malleable ways. Based on the stone tools and other objects made by anatomically modern people during the last tens of thousands of years, the products of language and of behavior mediated by language vary far more and in a much more intricate manner in time and space than any products found in the early Paleolithic record. The assumption that early hominid toolmaking was mediated by language, and that culture arrived wholesale when hominids began making tools, obscures the lengthy evolutionary history that the complex capacities underlying language and culture must have undergone during the course of human evolution.

Components of the Monolith of Culture

A significant part of the problem is that no accepted definition or concept of culture exists that takes into account the fact that humans have evolved. Culture tends to be

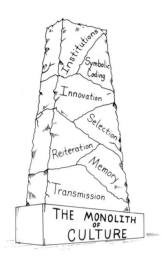

FIGURE 3.11

The "Monolith of Culture," and a new breakdown of the evolved components of cultural behavior. *Transmission* is a complex process that usually involves learning by observation or instruction. Transmitted information may be stored in order to be used (*memory*), and the tendency to reproduce or imitate stored behavior patterns or other transmitted information may be termed *reiteration*. New information may sometimes be produced (*innovation*) as a result of problem solving or individual variations in imitating transmitted information. Not all nongenetic information is retained or transmitted among individuals; *selection* entails processes of filtering new information by individuals and social groups. *Symbolic coding* refers to the encoding of information by using arbitrary and nonarbitrary symbols as in human language. By providing meaning, these symbols and their manipulation become ways to transmit and retain information. Finally, *institutionalization* consists of the processes whereby information and symbolic codes become organized into complex social order. These processes arise from the interplay between social organization and behavior and symbolic codes, and they result in the organizing of social relationships in complex ways.

treated in a monolithic way, as if **ethnologists** who discuss modern peoples, archeologists who discuss stone tools, and primatologists who discuss the learned behaviors of nonhuman primates are all talking about the same thing. Yet this is not the case. Based on what is now known about early hominid technology and the toolmaking and social behaviors of other primates, a revolutionary new view of culture seems to be needed.

Figure 3.11 suggests one possible analysis of the monolith of culture. According to this idea, culture is not an entity that is either present or absent. Rather, it consists of a set of components—aspects of behavior, cognition, and social interaction—all of which may be involved in how information held in an individual's **neurons** but not coded in the genes may be passed between individuals, used, and applied (Boyd and Richerson, Chapter 6). These components of culture are shown and described in Figure 3.11. When forged together, these components comprise culture in the modern human sense. Varying degrees of the bottom five elements in the monolith are also known to occur in the learned behavior of nonhuman primates, or at least have been inferred from observations of specific social groups of apes or monkeys. Intergroup differences in learned behavior have been detected in some species of nonhuman primates that are maintained beyond the lifetimes or experiences of single individuals (McGrew et al., 1979; McGrew, 1983).

During the early periods of hominid evolution there is little evidence to suggest that any but these most basic elements enabling learned behavior traditions had operated. The lack of change in archeologically detected behaviors strongly suggest the dominance of cognitive and social processes of reiteration over innovation. Only later, mainly during the late Pleistocene, do we see evidence in archeological remains of behavioral flexibility and innovation that is typical when symbolic codes are used to transmit information (for example, modern human linguistic capacities). Complex organization of societies based on institutions also may appear around the same time (by the **Upper Paleolithic**), creating heterogeneity in cultural behavior, as expressed in material remains, and the development of geographically and temporally distinctive styles of behavior (for example, in tool manufacture).

The problem in applying the term culture to the behavior of early hominids arises from the monolithic anthropological concept of culture used in reference to modern humans. Without clear reference to the evolved complex of components in Figure 3.11 the term culture helps very little to comprehend the origins of human behavior. This is because the monolith, as a whole, poorly represents the behaviors of Pliocene and early Pleistocene hominids, including tool manufacture and other very basic activities that can be distinguished archeologically. Certain aspects of the way modern people transmit cultural information appear to have been engaged by 2 million years ago, but not the entire monolith of culture.

■
CONCLUSIONS

Tool manufacture, habitation of home bases, and culture are often deemed to have comprised fundamental aspects in the adaptations of Pliocene and early Pleistocene toolmakers, critical to the success of the genus *Homo*. By attributing these aspects of behavior to the earliest stone toolmakers, paleoanthropologists have tended to treat these hominids very much like modern humans, specifically foragers (hunter-gatherers). This chapter has examined recent studies that contradict these traditional views. Tools, home bases, and culture remain important elements in the study of human origins. But new perspectives have arisen from a synthesis of (a) ideas from behavioral ecology, (b) detailed paleoecological study of early hominid sites, and (c) behavioral studies of nonhuman primates. The new perspectives developed in this chapter can be summarized as follows.

Resource Transport

According to what has long been a standard in anthropological thinking, the earliest production of stone tools signified a cardinal event in human evolution. It marked a formative shift in intelligence, manipulative skill, and other capacities fundamental to being human. Stone toolmaking greatly amplified hominid access to certain resources. But more important, it is the aggregated nature of Oldowan sites in a wider landscape context that stimulates a revised view of what the earliest known toolmaking ultimately meant to at least some populations of hominids between about 2.5 and 1.8 million years ago. Making tools out of natural materials, hauling them some distance from their source locations, and creating piles of debris at certain points on the landscape were all important antecedents developed later in the genus *Homo*. In contrast with traditional views, the movement of food and artifact resources may have been of far greater evolutionary consequence than the act of making simple tools itself. The dual transfer of stones and foods through the habitats of hominid toolmakers was a critical ingredient of the Oldowan, one that provided a distinctive selective advantage for at least certain early stone toolmakers (Potts, 1991).

Home Bases

By analogy to modern foragers and the activities of families in all modern human societies, it has long been thought that the earliest archeological sites were home bases. Study of the detailed taphonomic and behavioral evidence preserved at early archeological sites has now given rise to alternative models. Early toolmakers may have created accumulations of debris by repeatedly bringing food to stones left nearby (stone caches) or by repeatedly visiting stationary environmental resources (routed foraging). These alternate models suggest that the oldest hominid archeological sites were not necessarily areas of complex social interaction and food sharing, although this remains a possibility to be evaluated with future archeological research.

Criteria may be suggested that distinguish among these hypothetical models. The stone cache model requires evidence of long-term attraction of large carnivores, which are potential predators, to the remains of carcasses processed at sites by hominids, and by the presence of unmodified pieces of stone raw material brought from distant sources. The routed foraging model requires that archeological concentrations occurred at or adjacent to resources that had permanent locations on the landscape; evidence for such spatial correlations may be detectable using paleolandscape excavations of broad lateral scope. The central place foraging, or home base, model requires that conditions existed that fostered the aggregation of animal bones, other foods, and stone tools at the primary social focus of hominid groups. The occurrence of shelters, hearths, or other potential signals of social cohesion aid in discerning such campsites in the more recent archeological record.

In contrast with the idea that full-blown home bases originated with the Oldowan, very early archeological sites may represent a more fundamental solution to the problem of the distances that separated stone and food resources. We may now consider that this step created the initial conditions for later changes in the spatial grouping of hominids and the development of home bases.

Culture

We have also seen that the concept of culture in the modern human sense is inappropriate to apply to the oldest stone toolmakers. In order for the anthropological concept of culture to be consistent with the fact that humans have evolved, a new construction of the culture concept is needed. Archeologists have tended to equate the manufacture of tools, and the occurrence of tool assemblages, with culture. However, "Oldowan culture" or "Acheulian culture" may have involved the operation of rather different combinations of cognitive, behavioral, and social processes than is expressed in Upper Paleolithic toolmaking.

The earliest known hominid toolmaking may reflect little more than the application of learning capacities found widely in nonhuman primates to a novel activity, stone flaking. Applying a single, monolithic concept of culture to the stone tool record is a non-evolutionary approach to early hominids, for it draws attention away from alterations in the cognitive, behavioral, and social processes that created cultural behavior over time. The outmoded practice of calling stone tools "culture," and referring to the oldest stone tools as the "beginning of culture," likely will persist until a new formulation of cultural behavior taking human evolution into account is adopted.

Ecological Approaches

Evolution requires context, a history of conditions that affected the survival and reproduction of organisms. Investigations over the past 15 years have demonstrated that the detailed characteristics of archeological stone and faunal collections preserve crucial information about hominid activities, while other studies focusing on paleoenvironmental variables have provided significant data about the contexts of these activities. Information is available about spatial and temporal variations in vegetation, climate, stone sources for toolmaking, and carnivore activities (and their interaction with hominid activities at sites). Research has begun to expand from the archeology of individual, isolated sites to encompass widespread stratigraphic layers exposed over many kilometers to place evidence of hominid activities in a landscape context. Excavation and taphonomic analysis of factors underlying the preservation of buried materials is essential to all of these research efforts.

The paleoecological approach thus involves more than traditional archeology, integrating evidence from many fields of study. It encourages thinking outside the strictures of hominid fossil bones and stone tools. Although the latter are essential direct clues about hominids, these reflect dynamic interaction in a broader ecological context. Understanding the ecological context enables reseachers to assess the conditions under which certain hominid behaviors first gained expression, and how these departed from the survival strategies and evolutionary trajectories of contemporaneous animals. The conditions of extinction of hominid species will be just as important to investigate as the conditions of survival. These paleoecological objectives will be met by careful and innovative research on fossil animals, environmental clues, and the traces of hominid behaviors. Direct investigation of the contents of the geologic record is the one avenue that exists to avoid simple extrapolations from the modern human present to these early hominids that we so strongly wish to understand.

Acknowledgments

Research support from the following institutions and funding agencies is gratefully acknowledged: Smithsonian Institution, National Museums of Kenya, National Museum of Tanzania, National Science Foundation, Harvard University, and the Boise Fund (University of Oxford). Illustrations and other invaluable assistance with the manuscript were furnished by Jennifer Clark.

■

SUGGESTED READINGS

Foley, R. 1987. *Another Unique Species* (London: Longman Scientific and Technical).

Isaac, G. L. 1984. The archaeology of human origins: Studies of the Lower Pleistocene in East Africa 1971–1981. *Advances in World Archaeology 3*: 1–87.

Potts, R. 1988. *Early Hominid Activities at Olduvai* (New York: Aldine de Gruyter).

Potts, R. 1989a. Ecological context and explanations of hominid evolution. *Ossa 14*: 99–112.

Toth, N., and Schick, K. D. 1986. The first million years: The archaeology of protohuman culture. In: *Advances in Archaeological Method and Theory 9*: 1–96.

■

REFERENCES

Andrews, P. 1989. Palaeoecology of Laetoli. *Journal of Human Evolution 18*: 173–181.

Binford, L. R. 1984. *Faunal Remains from Klasies River Mouth* (Orlando: Academic Press).

Blumenschine, R. J. 1986. Early hominid scavenging opportunities: Implications of carcass availability in the Serengeti and Ngorongoro ecosystems. *British Archaeological Reports International Series 283*.

Blumenschine, R. J. 1987. Characteristics of an early hominid scavenging niche. *Current Anthropology 28*: 383–407.

Blumenschine, R. J., and Masao, R. 1991. Living sites at Olduvai Gorge, Tanzania? Preliminary landscape archaeology results in the basal Bed II lake margin zone. *Journal of Human Evolution 21*: 451–462.

Blumenschine, R. J., and Selvaggio, M. M. 1988. Percussion marks on bone surfaces as a new diagnostic of hominid behavior. *Nature 333*: 763–765.

Boesch, C., and Boesch, H. 1984. Mental map in wild chimpanzees: An analysis of hammer transports for nut cracking. *Primates 25*: 160–170.

Bonnefille, R. 1985. Evolution of continental vegetation: The palaeobotanical record from East Africa. *South African Journal of Science 81*: 267–270.

Bonnefille, R., and Riollet, G. 1980. Palynologie, vegetation et climats de Bed I et Bed II a Olduvai, Tanzania. *Proceedings of the Eighth PanAfrican Congress of Prehistoric and Quaternary Studies*, September 1977, Nairobi, pp. 123–127.

Brain, C. K. 1981. *The Hunters or the Hunted* (Chicago: Univ. of Chicago Press).

Brain, C. K. 1985. Cultural and taphonomic comparisons of hominids from Swartkrans and Sterkfontein. In: E. Delson (Ed.), *Ancestors: The Hard Evidence* (New York: Alan R. Liss), pp. 72–75.

Bunn, H. T. 1989. Diagnosing Plio-Pleistocene hominid activity with bone fracture evidence. In: R. B. Bonnichsen and M. H. Sorg (Eds.), *Bone Modification* (Orono, Maine: Center for the Study of Early Man), pp. 299–315.

Cerling, T., and Hay, R. L. 1986. An isotopic study of paleosol carbonates from Olduvai Gorge. *Quaternary Research 25*: 63–78.

Dart, R. A. 1955. Cultural status of the South African Man-apes. *Annual Report of the Smithsonian Institution*, pp. 317–338.

Dart, R. A. 1957. The Osteodontokeratic Culture of *Australopithecus prometheus*. *Transvaal Museum Memoirs*, no. 10.

Deino, A., and Potts, R. 1990. Single crystal ^{40}Ar/^{39}Ar dating of the Olorgesailie Formation, southern Kenya rift. *Journal of Geophysical Research 95* (B6): 8453–8470.

Galton, P. 1985. Diet of prosauropod dinosaurs from the late Triassic and early Jurassic. *Lethaia 18*: 105–123.

Goodall, J. 1986. *The Chimpanzees of Gombe* (Cambridge: Harvard Univ. Press).

Harris, J. W. K. 1983. Cultural beginnings: Plio-Pleistocene archaeological occurrences from the Afar, Ethiopia. *African Archaeological Review 1*: 3–31.

Hay, R. L. 1976. *Geology of the Olduvai Gorge* (Berkeley: Univ. of California Press).

Isaac, G. L. 1972. Some experiments in quantitative methods for characterizing assemblages of Acheulian artifacts. *VIième Congrès Panafricain de Préhistoire et de l'Etude du Quaternaire, 1967* (Chambéry: Imprimeries Réunies de Chambéry).

Isaac, G. L. 1977. *Olorgesailie: Archaeological Studies of a Middle Pleistocene Lake Basin* (Chicago: Univ. of Chicago Press).

Isaac, G. L. 1978. The food sharing behavior of protohuman hominids. *Scientific American 238*: 90–108.

Isaac, G. L. 1983. Some archaeological contributions towards understanding human evolution. *Canadian Journal of Anthropology 3*: 233–243.

Isaac, G. L. 1984. The archaeology of human origins: Studies of the Lower Pleistocene in East Africa 1971–1981. *Advances in World Archaeology 3*: 1–87.

Keeley, L. 1980. *Experimental Determination of Stone Tool Uses* (Chicago: Univ. of Chicago Press).

Keeley, L., and Toth, N. 1981. Microwear polishes on early stone tools from Koobi Fora, Kenya. *Nature 293*: 464–465.

Krebs, J. R., and Davies, N. B. 1991. *Behavioural Ecology: An Evolutionary Approach*, 3rd ed. (Oxford: Blackwell).

Laporte, L. F., and Zihlman, A. L. 1983. Plates, climate, and hominoid evolution. *South African Journal of Science 79*: 96–110.

Leakey, L. S. B. 1935. *The Stone Age Races of Kenya* (London: Oxford Univ. Press).

Leakey, L. S. B., Tobias, P. V., and Napier, J. R. 1964. A new species of the genus *Homo* from Olduvai Gorge. *Nature 202*: 7–9.

Leakey, M. D. 1971. *Olduvai Gorge*, Vol. 3 (London: Cambridge Univ. Press).

McGrew, W. C. 1983. Animal foods in the diets of wild chimpanzees (*Pan troglodytes*): Why cross-cultural variation? *Journal of Ethology 1*: 46–61.

McGrew, W. C. 1989. Why is ape tool use so confusing? In: V. Standen and R. A. Foley (Eds.), *Comparative Socioecology: The Behavioural Ecology of Humans and Other Mammals* (Oxford: Blackwell), pp. 457–472.

McGrew, W. C., Tutin, C. E. G., and Baldwin, P. J. 1979. Chimpanzees, tools, and termites: Cross-cultural comparisons of Senegal, Tanzania, and Rio Muni. *Man 14*: 185–214.

Oakley, K. P. 1961. *Man the Toolmaker* (Chicago: Univ. of Chicago Press).

Pianka, E. R. 1983. *Evolutionary Ecology* (New York: Harper and Row).

Potts, R. 1986. Temporal span of bone accumulations at Olduvai Gorge and implications for early hominid foraging behavior. *Paleobiology 12*: 25–31.

Potts, R. 1988. *Early Hominid Activities at Olduvai* (New York: Aldine de Gruyter).

Potts, R. 1989b. Olorgesailie: New excavations and findings in Early and Middle Pleistocene contexts, southern Kenya rift valley. *Journal of Human Evolution 18*: 477–484.

Potts, R. 1991. Why the Oldowan? Plio-Pleistocene toolmaking and the transport of resources. *Journal of Anthropological Research 47*: 153–176.

Roughgarden, J., May, R. M., and Levin, S. (Eds.). 1989. *Perspectives in Ecological Theory* (Princeton: Princeton Univ. Press).

Rubenstein, D. I., and Wrangham, R. W. (Eds.). 1986. *Ecological Aspects of Social Evolution: Birds and Mammals* (Princeton: Princeton Univ. Press).

Schick, K. D. 1987. Modeling the formation of early stone age artifact concentrations. *Journal of Human Evolution 16*: 789–807.

Susman, R. L. 1988. Hand of *Paranthropus robustus* from Member 1, Swartkrans: Fossil evidence for tool behavior. *Science 240*: 781–784.

Susman, R. L. 1991. Who made the Oldowan tools? Fossil evidence for tool behavior in Plio-Pleistocene hominids. *Journal of Anthropological Research 47*: 129–151.

Toth, N. 1985. The Oldowan reassessed: A close look at early stone artifacts. *Journal of Archaeological Science 12*: 101–120.

Toth, N. 1987. Behavioral inferences from early stone artifact assemblages: An experimental model. *Journal of Human Evolution 16*: 763–787.

Toth, N., and Schick, K. D. 1986. The first million years: The archaeology of protohuman culture. In: *Advances in Archaeological Method and Theory 9*: 1–96.

Van Couvering, J. A. H. 1980. Community evolution in East Africa during the late Cenozoic. In: A. K. Behrensmeyer and A. Hill (Eds.), *Fossils in the Making* (Chicago: Univ. of Chicago Press), pp. 272–298.

Washburn, S. L. 1960. Tools and human evolution. *Scientific American, 203(3)*: 3–15.

Winterhalder, B. 1980. Environmental analysis in human evolution and adaptation research. *Human Ecology 8*: 135–170.

CHAPTER 4

NEW VIEWS ON MODERN HUMAN ORIGINS

■

Christopher B. Stringer*

■

INTRODUCTION

Our species, *Homo sapiens*, has reached nearly every part of the Earth's surface, and has now reached out into space to visit the moon. But 2 million years ago, our ancestors were apparently restricted to only one continent—their continent of origin, Africa. The process by which we became the most widespread mammalian species really began about 1 million years ago during the early Pleistocene, when the species *Homo erectus* began to disperse out of its African homeland into Asia, and from there into Europe and Southeast Asia. Initially, and for a considerable time subsequent to this, they kept to the warmer zones of the world, zones that were nevertheless not climatically stable, particularly from about 700,000 years ago, when the orbital cycles of the Earth progressively switched to the production of climatic extremes of ice and aridity followed by shorter intervals about every 100,000 years of warmer and moister conditions. *Homo erectus* continued to evolve during the middle Pleistocene in the four primary geographic regions for which a reasonable fossil record or archeological record exists (Africa, Europe, eastern Asia, and the Malay Archipelago, then mainly a contiguous land mass connected to southern Asia). By 300,000 years ago, *Homo erectus* had changed sufficiently that most paleoanthropologists recognize that one or more new species had appeared. However, it is possible that surviving groups of *Homo erectus* persisted in the Far East and particularly in the Malay Archipelago for some considerable time after the appearance of new species elsewhere (for example, represented by the Javan Ngandong or Solo sample). Figure 4.1 shows a selection of important sites for fossil human remains for the last million years, arranged by geographical origin and possible or probable date.

The evolutionary changes of the middle Pleistocene were primarily physical rather than behavioral, since it appears that cultural evolution as reflected by changes in

*Human Origins Group, Department of Palaeontology, The Natural History Museum, London SW7 5BD

FIGURE 4.1

Fossil hominid sites of the last million years are widespread geographically, unlike the situation before 1 million years ago, when all known fossil hominids occur in a restricted number of sites in eastern and southern Africa. This diagram shows selected hominid-bearing sites arranged by continent and by probable or possible age, shown in thousands of years on the left.

Age*10³	Africa	Eurasia	E. Asia / Australasia
30	AFALOU	CRO-MAGNON MLADEC ST.CÉSAIRE KSAR AKIL LeMOUSTIER	ZHOUKOUDIAN MUNGO, WLH-50 NIAH
50		AMUD, LA CHAPELLE KEBARA, GUATTARI	
100	BORDER CAVE KLASIES, GUOMDE OMO KIBISH IRHOUD, NGALOBA	SKHUL QAFZEH TABUN, SACCOPASTORE KRAPINA BIACHE EHRINGSDORF SWANSCOMBE	XUJIAYAO NGANDONG MABA DALI
	BROKEN HILL		
250	 ELANDSFONTEIN, NDUTU BODO, SALÉ	ATAPUERCA PETRALONA BILZINGSLEBEN ARAGO	JINNIU SHAN HEXIAN & NGAWI YUNXIAN ZHOUKOUDIAN & SAMBUNGMACHAN
500	BARINGO OLDUVAI, TERNIFINE	MAUER	 LANTIAN, SANGIRAN
1000	OLDUVAI	DMANISI	

stone tool technology was painfully slow for most of the Paleolithic (see Potts's discussion in Chapter 3). The physical changes were mainly those of a significant increase in brain size, a slight reduction in robustness in known parts of the skeleton, and some realignments of the face and brain case. These more derived human populations are often grouped under the term **archaic Homo sapiens**, implying that they are the earliest members of our own species but are not yet modern. However, it is not at all clear that this is the most appropriate classification for them (Figure 4.2). Usually included within this blanket term of *archaic sapiens* are the **Neanderthals**, who occupy a special place both for scientists and the general public as the archetypal early humans. This is because they are the best known primitive humans. For the public they are regular inhabitants of cartoons; for the scientists they were the first early human type to be identified from fossils, they provide the largest sample of archaic human remains, and they have been studied for longer and in greater detail than any other early hominids.

MODELS OF MODERN HUMAN ORIGINS

It is generally agreed that people who closely matched the physical and behavioral patterns of living humans appeared relatively late in the Pleistocene. Several distinctly human behavioral attributes, such as the production of carved and painted images (Conkey, Chapter 5), may have emerged as recently as 40,000 years ago, and modern physical attributes appeared within the last 150,000 years. Despite this general agreement on when the changes occurred there is no agreement as to the processes that produced the changes that led to our evolution. In one view, that of the *multiregional model*, the dispersal of *Homo erectus* out of Africa and into Europe, Asia, and the Malay Archipelago initiated the evolutionary radiation of modern

FIGURE 4.2

A possible classification of fossil hominids superimposed upon those listed in Figure 4.1. A = *Homo erectus*; B = *Homo heidelbergensis* (or, along with C, a widely defined *Homo neanderthalensis*); C = *Homo neanderthalensis*; D = *Homo sapiens*. B and C together have generally been recognized as "archaic *Homo sapiens*."

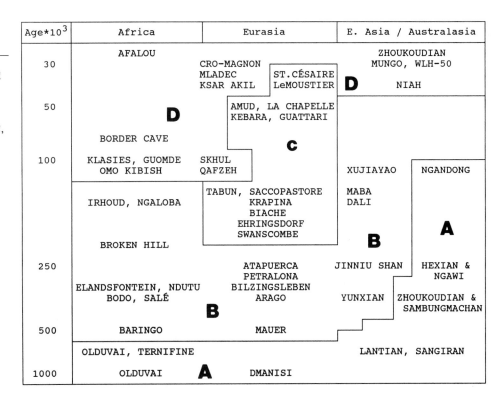

people in the Old World (Wolpoff et al., 1984; Wolpoff, 1989). Human populations across the world shared a common **gene pool**, so that there was no further evolutionary splitting during the last million years. **Anatomically modern *Homo sapiens*** gradually developed everywhere, through both local evolution and through **gene flow** between regions. There would have been no single center for the origin of modern people because the characters combined in modern people today had different geographical origins and have been spread by mate exchanges and population movements (but not by large scale migrations or replacements). The evolution of modern humans would have been a process covering the whole of the last million years or so, although the combination of modern characters led to a so-called crystallization of the modern morphology during the last 100,000 years. In fact, since taxonomic boundaries would be expected to be unclear under this model, the most reasonable classification for the populations of the last 1 million years would be to include them all within our own species, *Homo sapiens*. However, evolving regional populations also developed and maintained distinctive characteristics through local adaptation, and these persist through time to demonstrate *regional continuity*. Examples of these are the prominent noses of Europeans, present in Neanderthals 120,000 years ago, and retained by modern Europeans, the flat faces and prominent cheek bones of the East Asians, present in *Homo erectus* at Zhoukoudian (Peking Man) 300,000 years ago and retained in modern oriental populations, and the narrower, flatter foreheads of Javan *Homo erectus* 700,000 years ago, retained in modern native Australians. A representation of the multiregional model is shown in Figure 4.3. Here many of the regional characteristics of modern humans mainly developed during the middle Pleistocene.

A related model to that of multiregionalism is the *gene flow/assimilation model*, which suggests that gene flow between the different regional populations of early humans might not have been equal or persistent through time and space (Smith et

FIGURE 4.3

Graphic representation of the *multiregional model* of modern human origins, superimposed on the fossil hominids listed in Figure 4.1. Vertically directed open arrows indicate continuity of regional traits through time; horizontal dark arrows represent gene flow between regions that allowed traits originating in one area to spread elsewhere. All hominids in this chart would be classified as *Homo sapiens* under the multiregional model, or would be arbitrarily divided into older *Homo erectus* and younger *Homo sapiens* at some point in the evolutionary continuum.

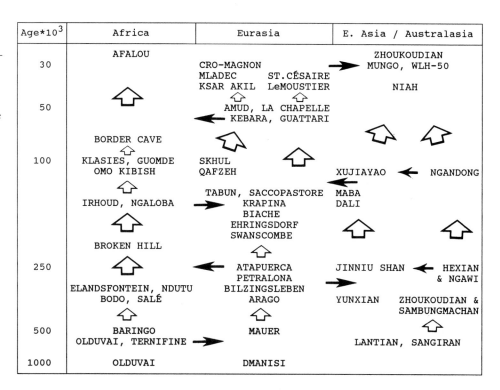

al., 1989). The archaic populations of *Homo erectus* or *Homo sapiens* might not have made equal contributions to the evolution of modern *Homo sapiens*, so more modern features could have evolved in certain areas such as Africa or the Middle East and then spread from there by the gradual intermixture of populations. Thus the evolution of modern humans could have been due to a blending of modern characters derived from a few centers of origin with local characteristics. Thus the Neanderthals of western Europe perhaps contributed little or nothing to the evolution of modern Europeans, but those of central Europe and the Middle East were perhaps gradually transformed by gene flow and selection favoring modern characters to become early modern people, showing a mixture of persisting Neanderthal features and more intrusive modern features. From this model, regional characteristics in each inhabited area could have blended with modern ones during the early late Pleistocene, 130,000 to 50,000 years ago, resulting in present-day human variation.

A third view of recent human evolution is that there was only one region where there was an evolutionary transition from *Homo erectus* through to modern *Homo sapiens*, and modern humans then spread from that center of origin to the rest of the inhabited world by dispersal (Stringer and Andrews, 1988; Stringer, 1989, 1990b, 1992). Although there could have been a limited amount of hybridization with the descendants of *Homo erectus* in other regions of the world, this had little impact on the physical appearance of early modern people in each region. A racially undifferentiated stock of early modern people spread to each region of the world and only then started to develop the distinctive characters found in that region today. This monogenetic view of human evolution has been suggested from time to time in the past, but there was previously insufficient evidence to indicate where this original homeland of *Homo sapiens* might have been (Howells, 1976; Beaumont et al., 1978). However, in the last few years, the evidence of fossils and of genetic studies of recent human samples have been combined to support an African origin for modern *Homo sapiens* (Figure 4.4).

FIGURE 4.4

Graphic representation of the *"Out of Africa" model* of modern human origins, superimposed on the fossil hominids listed in Figure 4.1. Modern humans originated in Africa, probably between 100,000 and 150,000 years ago from ancestors like Jebel Irhoud. From that single source modern humans then spread throughout the Old World (arrows) with little or no genetic input from the resident archaic populations, which became extinct (below the solid line). Archaic populations (Neanderthals) persisted in Europe until less than 40,000 years ago.

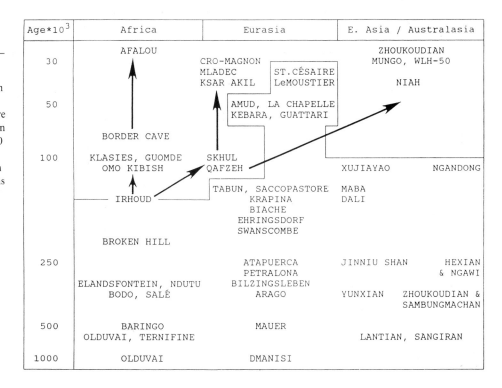

The emergence of the "Out of Africa" model has greatly sharpened the debate about modern human origins, leading to an increasing polarization of viewpoints in the face of heightened public and media interest, and a spate of scientific conferences and specialist and popular publications (e.g., Mellars and Stringer, 1989; Trinkaus, 1989; Bräuer and Smith, 1992; Brown, 1990; Fagan, 1990). In addition to proponents of a strict regional continuity model (Wolpoff and Thorne, 1984) or a strict "Out of Africa" model (Stringer and Andrews, 1988), there are others who hold a range of intermediate positions, envisioning either a very limited amount of genetic input from archaic populations through to substantial input from at least some of these (Bräuer, 1992; Smith and Trinkaus, 1991). At this exciting stage of the discussions about modern human origins I review some of the recently available evidence and look at how this can be interpreted. I conclude with an overview of how I think we have progressed toward resolving some of the outstanding questions about the origins of our own species.

THE FOSSIL EVIDENCE

The fossil record provides direct clues pertaining to changes in skeletal anatomy that have occurred during the emergence of anatomically modern humans. In order to interpret what these changes mean, it is first necessary to consider the anatomical differences between undisputed representatives of archaic *Homo sapiens* or *Homo erectus* on the one hand, and anatomically modern humans on the other (Figure 4.5). Modern populations are characterized by a combination of attributes that include decreased robustness of the postcranium; thinner bones of the brain case; higher foreheads and reduction in the size of brow ridges; shorter and rounder brain cases rather than long and low ones; relatively narrow cranial base; presence of distinct

FIGURE 4.5

Comparison of the cranium of *Homo erectus* (left), anatomically modern *Homo sapiens* (center) and Neanderthal (right) viewed from back, front and side. The skull of *Homo erectus* is Sangiran 17 from the Kabuh Formation, Java. The anatomically modern specimen is from a recent population in Indonesia. The Neanderthal specimen is the La Ferrassie I skull, found in a French rock shelter. Photos by the Photographic Unit, Natural History Museum, London.

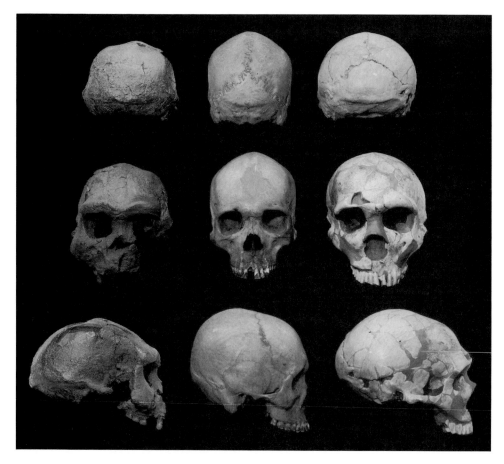

chins; and often, increased cranial capacity. Neanderthals share some features with anatomically modern humans, such as a large cranial capacity, but they retain the archaic features of prominent brow ridges, more sloping foreheads, long, low brain cases, lack of a prominent chin, and skeletal robustness. In addition, Neanderthals exhibit their own distinctive specializations, such as an extremely broad nasal cavity, a characteristic occipital "bun" at the back of the head, and marked prominence, or forward projection, of the mid-facial region.

As the fossil evidence continues to accumulate at a steady pace, many of the new discoveries seem mainly to reinforce existing knowledge. Occasionally, however, there is a dramatic breakthrough that may fill an existing gap in the evidence. These gaps may be geographical, chronological, or morphological (anatomical). An example of a find that filled all three gaps concerned the period between 100,000 and 300,000 years ago in the Far East. One fairly complete skull was known from the Chinese site of Dali (Figure 4.6), believed to date to about 150,000 years ago, but other finds from this time were rather fragmentary. However, in 1984, a partial skeleton was excavated from a fissure filling at Jinniu Shan in northeast China which Chinese scientists believe may be as old as 250,000 years (Lu, 1990; Chen and Yinyun, 1991; Pope, 1992). If the age estimates are correct, then this individual may have lived close in time to the last examples of *Homo erectus* in China. However, this was clearly not a specimen of *Homo erectus* despite its robust skeleton and large brow ridges. The Jinniu Shan individual had a modern-sized brain (about 1,300 cm³) housed in a thin-walled brain case. Although they differ from each other in a number of respects, both the Jinniu Shan and Dali specimens are quite distinct from *Homo erectus* and share advanced features with later humans, including modern humans.

FIGURE 4.6

The Dali cranium, excavated from a sandy gravel layer along a tributary of the Huang River in central China, and found in association with stone tools made of flint and quartzite. It is often referred to as an archaic *Homo sapiens*. The lower part of the face has been distorted by upward crushing. Photo by C. Stringer, courtesy of Dr. X. Z. Wu.

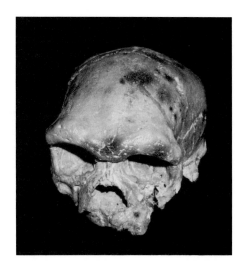

They show that China was not a backwater in the story of human evolution and that there was probably not a straightforward evolutionary sequence from *Homo erectus* to *Homo sapiens* in the area (Stringer, 1990a).

An example of a morphological gap was our lack of knowledge of certain parts of the Neanderthal skeleton. There were numerous partial Neanderthal skeletons from Europe and the Middle East, and yet no single individual had a nearly complete trunk or pelvic skeleton. This gap was filled by a discovery in the Kebara cave in Israel in 1983. Here, a Neanderthal skeleton was excavated that was virtually complete from the lower jaw and **hyoid** (throat) bone down to the hip joints. It was from a large and heavily built male individual who was apparently deliberately buried, although the skull had seemingly been taken out of the grave afterwards by humans or by animal scavengers (Bar-Yosef et al., 1988). The find has provided important new data on the body form of Neanderthals. For example, the hyoid bone is the first known for any early fossil hominid and bears a close resemblance to those of modern *Homo sapiens* (what this may or may not signify in terms of Neanderthal vocal abilities is a matter of considerable dispute).

A third example, of a chronological gap being filled, concerns our previous ignorance of what the last Neanderthals looked like. Were they as distinctive from modern *Homo sapiens* as earlier examples, or were they approaching a modern anatomy, forming an evolutionary sequence toward modern humans or reflecting the influx of modern genes (gene flow) from elsewhere? This gap has started to be filled by discoveries in France and Croatia. The French discovery was of a Neanderthal skeleton in a **Châtelperronian** archaeological level at the rock shelter of Saint-Césaire (Lévêque and Vandermeersch, 1981). The Châtelperronian is a stone tool industry seemingly more advanced than, but closely related to, the **Middle Paleolithic** industries of the Neanderthals. This discovery seemed to confirm the coexistence of Neanderthals and early anatomically modern humans in Europe (**Cro-Magnons**) between 30,000 and 40,000 years ago, because the Châtelperronian tools had already been **radiocarbon dated** from other sites to this time period. Sites with a different type of stone tool industry (the **Aurignacian**) were also known from the same time period in Europe, but these sites had produced remains of Cro-Magnons. The Croatian site of Vindija also has fossils of what are believed to be late Neanderthals, although these are not so well dated nor as individually well preserved as at Saint-Césaire (Wolpoff et al., 1981). The Saint-Césaire individual has a fairly complete face and lower jaw, and these are very like those of earlier Neanderthals (Figure 4.7). Moreover the forehead

FIGURE 4.7

Skull demonstrating classic Neanderthal structure; note especially the low forehead, heavy brow ridges, broad nasal aperture, and receding cheek bones. This male specimen is from the rock shelter of La Ferrassie, France, where remains were also found of an adult female, a child, two infants, a neonate, and two fetuses. The hominid fossils occurred in archeological association with **Mousterian** stone tools, and fossil mammoth, hyena, red deer, wild ox, pig, and horse. Photo by C. Stringer, courtesy of Musée de l'Homme, Paris.

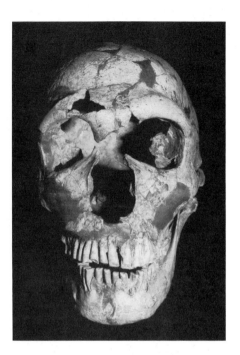

is low, the brow ridges are strong and prominent, and the face has the midfacial and nasal prominence of other Neanderthals. The Vindija remains are Neanderthal-like in most respects but, like Saint-Césaire, have relatively narrower noses. However, here the brow ridges appear to be weaker, especially at the sides, and it has been estimated that the middle of the Vindija faces was less prominent. Interpretation of the Saint Césaire and Vindija finds has been a source of much debate.

THE DATING EVIDENCE

The ages of various fossils have important bearing on assessing the different models of modern human origins. Did anatomically modern populations occur earlier in some areas than others? Did two anatomically distinctive populations exist at the same time side by side, thus implying limited or no genetic admixture? In many cases, reliable interpretation of population movements or evolutionary changes depends on accurate dates. Dates can sometimes be obtained from chemical analyses of the fossils themselves, and with other methods, dates can be obtained from associated materials that occur in the same stratigraphic level as the fossils in question.

Dating techniques have been developed or refined over the last decade, allowing a greater precision of dating for methods such as radiocarbon and **uranium series dating,** or a wider time range of dating for methods such as **thermoluminescence** and **electron spin resonance** (Aitken, 1990). The latter two methods have had a special impact, particularly on our understanding of later human evolution in the Middle East (Stringer 1988; Grün and Stringer, 1991). For example, thermoluminescence has been used to date burned flints, and electron spin resonance dates animal tooth fragments that come from the stratigraphic level of the Kebara Neanderthal; these analyses indicate an age of about 60,000 years (Valladas et al., 1987; Schwarcz et al., 1989). This was a useful confirmation of the expected antiquity of the individual. However, when these same techniques were applied to early anatomically modern

FIGURE 4.8

Entrances of two limestone caves in Israel that have yielded some of the earliest fossils of modern *Homo sapiens*, recovered during archeological excavation of the stratigraphic layers preserved on the cave floors. The shallow cave or rock shelter at Skhul (left), on the western slope of Mount Carmel, contained the remains of at least ten individual hominids, 7 adults and 3 youngsters. A larger cave — 20 m broad and 12 m deep beyond the relatively narrow opening — on the flank of Mount Qafzeh (right) contained skeletons of 8 adults, 2 infants, and 2 children of about 3 and 10 years. At both sites, a rich diversity of Middle Paleolithic tools was found, and also other vertebrate fossils including hippopotamus, rhinoceros, wild ox, wild horse, gazelle, and several species of deer, reflecting the different climate and biota of the Middle East during the late Pleistocene. Photos by C. Stringer.

human remains from the cave sites of Skhul and Qafzeh, also in Israel (Figure 4.8; Vandermeersch, 1981), there was great surprise when they were dated to about 80,000 to 120,000 years old, twice the expected age (Valladas et al., 1988; Schwarcz et al., 1988; Stringer et al., 1989). In other words, these early modern humans did not succeed the Neanderthal from Kebara, but apparently preceded him by a considerable period of time!

Similar dating work has demonstrated that Neanderthals probably persisted in the Middle East until at least 50,000 years ago (at the site of Amud Cave, Israel) and has also confirmed an even younger age for Neanderthals at the French site of Le Moustier (less than 40,000 years) and at Saint-Césaire (about 36,000 years; Grün and Stringer, 1991; Mercier et al., 1991). As we have already indicated it seems very probable that the last Neanderthals overlapped in time with early modern Cro-Magnons in Europe, which are dated to the period between 30,000 and 36,000 years ago, based on radiocarbon dates of the fossils themselves (at the sites of Kent's Cavern and Hahnöfersand) or of associated material (the sites of Cro-Magnon, La Crouzade, Vogelherd, Mladec, and Velika Pecina; Stringer and Grün, 1991).

Neanderthals related to those in Europe therefore persisted in the Middle East during the last 100,000 years, after the earliest known appearance of anatomically modern humans in the area (Figure 4.9). However, Neanderthals seemingly never appeared in Africa, where the populations of that continent had evolved from *Homo erectus* in a somewhat different manner. They did not develop all the Neanderthal specializations of the skull and body despite an overall general resemblance. In various sites scattered around the continent there is evidence of populations that were becoming more modern in their anatomy. The brain case was becoming higher, shorter, and more rounded, the brow ridges smaller, and the lower jaw lighter in build, with signs of a chin. The rest of the body is poorly known from African fossils

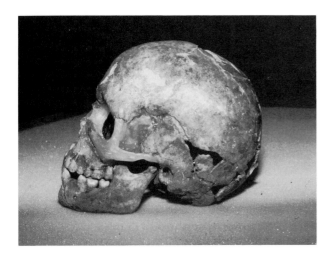 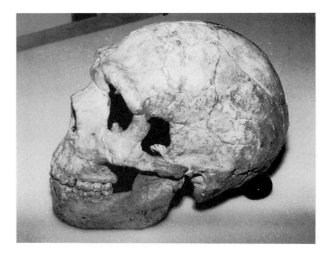

FIGURE 4.9

Skulls of anatomically modern human and Neanderthal from the Middle East. Left: Qafzeh XI, skull of a child, about 10 years of age, who along with the other Qafzeh hominids probably lived about 100,000 years ago, making them among the earliest known anatomically modern humans; the very modern appearance of Qafzeh XI is partially related to its youthfulness, but fully adult skulls from the same stratigraphic levels also have the appearance of modern humans rather than of Neanderthals (Figure 4.12). Right: Amud I, a Neanderthal skull from a cave along the Wadi Amud, Israel. With an estimated age of 50,000 years, the Amud hominid is the most recent dated Neanderthal from the Middle East. Photos by C. Stringer, courtesy of Dr. J. Zias.

of this date but the available evidence suggests that these populations were, like the Neanderthals, strongly built by modern standards, even if rather different in body proportions. It has recently been feasible to estimate the age of some of these possible African ancestors of modern humans. Two sites in northern Africa, Jebel Irhoud in Morocco, and Singa in the Sudan, have fossil material that shows a mixture of archaic and modern features (Figure 4.10). Until recent years both were very uncertainly dated, with radiocarbon ages on unassociated materials suggesting an antiquity of less than 40,000 years in each case. However, electron spin resonance estimates from associated animal teeth indicate a probable age of 90,000 to 190,000 years for both the Moroccan and the Sudanese sites (Grün and Stringer, 1991). Similar ages have been proposed for a number of other African localities with comparable fossils, such as Florisbad and the Cave of Hearths (South Africa), Ngaloba (Laetoli, Tanzania), Guomde (East Turkana, Kenya) and Omo Kibish (Ethiopia) (Klein, 1989; Bräuer, 1992).

Dating the earliest modern humans in Africa has also been difficult, but much progress has been made recently. Klasies River Mouth Caves in South Africa have produced a series of fragmentary cranial, dental, and postcranial remains, many of which are recognizably modern in morphology (Singer and Wymer, 1982; Deacon, 1989; Rightmire and Deacon, 1991). Using a series of different techniques it has been established that many of these date from 70,000 to 120,000 years ago (Deacon, 1989; Grün et al., 1990b). Further South African sites with modern fossils that are apparently of similar ancient age are the Die Kelders, Equus, and Border Caves. The latter site has produced a fragmentary skull and postcranial bones, a child's skeleton, and two lower jaws, all of which are modern in morphology (Beaumont et al., 1978). Some, at least, can be dated by electron spin resonance to more than 60,000 years old (Grün et al., 1990a). One early modern specimen, perhaps the most ancient of all,

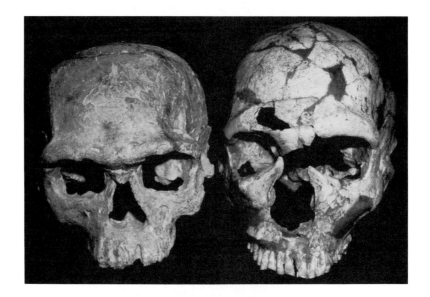

FIGURE 4.10

An African cranium of mixed archaic and modern features (left) compared with a much younger Neanderthal cranium from Europe (right). The African cranium is Irhoud I, preserved in the natural clay fillings of a fissured limestone deposit; the specimen was discovered during excavations at a barite mine, Jebel Irhoud, Morocco. The fissure fillings also contained abundant flint tools and a diversity of mammal fossils. Note the relatively high forehead, reduced brow ridges, narrow nasal aperture, and delicate cheek bones when compared with a Neanderthal skull, La Ferrassie I. Photos by C. Stringer, courtesy of the Musée de l'Homme, Paris.

has yet to have its suggested antiquity of 130,000 years confirmed by the new dating techniques. This is the partial skeleton found at Omo Kibish, Ethiopia, in 1967 (Leakey et al., 1969; Butzer, 1976; Day and Stringer, 1991; Day et al., 1991).

The ages of late archaic and early modern fossils in the Far East are less clear in most cases. Radiocarbon and thermoluminescence dates suggest that early modern people were already established in China (the Upper Cave at Zhoukoudian), Borneo (Niah Cave) and Australia (Lake Mungo) between 25,000 and 40,000 years ago (Kennedy, 1979; Klein, 1989; Stringer, 1990a). However, it is not known if they were present earlier than this. The antiquity of the last archaic populations in the Far East is also not certain. Only fragmentary Chinese fossils succeed Jinniu Shan and Dali, around 100,000 years ago, while in Java (Indonesia) it appears that the lineage of *Homo erectus* continued to a similar time period in the form of the Ngandong (Solo) fossil sample, mostly consisting of brain cases without faces or jaws (Santa Luca, 1980; Bartstra et al., 1988). The role of these Javan hominids in the evolution of modern humans in southeast Asia and Australia is hotly disputed (Wolpoff et al., 1984; Smith et al., 1989; Wolpoff, 1989; Stringer, 1990a).

THE MOLECULAR EVIDENCE

In addition to the morphological and temporal information about evolution obtained from the incomplete fossil record, another independent line of evidence is available when the species under study is still alive today: molecular **genetics**. The amount of molecular variability within a species, and the pattern of how that variation is geographically distributed on different land masses, should provide clues as to the relationships among human populations, and perhaps also the age and location of the ancestral source. Populations that are very recently descended from a common ancestor have not had the time to accrue as many **mutations** as what might be observed in long-lived species. Similarly, geographic areas with greater genetic diversity are more likely to have a long history of endemic evolution than areas with little molecular variation, all other things being equal.

Since 1986 the impact of molecular reconstructions of recent human evolution has grown (Cann et al., 1987; Cavalli-Sforza et al., 1988). These have sometimes been based on small data sets such as for proteins called **beta globins** or for small segments of the separately inherited molecules of **mitochondrial DNA** (mtDNA). However, samples sizes and resolution of the data have grown enormously since 1986, particularly since the advent of the **polymerase chain reaction** which allows small quantities of genetic material to be replicated in enormous quantities.

It is useful to distinguish the mtDNA results from those based on nuclear DNA or the molecules produced from nuclear DNA (such as beta globins). The latter comprise by far the larger part of the genetic data bases. The results from the mtDNA work have usually been presented in the form of **cladograms** of individual lineages (based on individual samples from populations), because mtDNA is passed exclusively through the female line and does not undergo **meiosis** and sexual reproduction. Each line of mtDNA thus represents, in effect, a separate evolutionary lineage. The results obtained from analysis of nuclear DNA, which does undergo **recombination**, have usually been presented as genetic distance trees, which are based on the overall patterns of genetic difference measured within and between populations, patterns that may be influenced by many population genetic processes besides simply the splitting of lineages and accumulation of variation in each one.

The original assumption that human mtDNA was strictly maternally inherited may prove to be incorrect (Gyllensten et al., 1991), but the analysis of individual lineages has provided both a distinctive evolutionary perspective and the potential for establishing a realistic time scale for the establishment of the observed differences among present-day lineages. Estimates for the origin of modern human mtDNA variation can be calculated based on assumptions about mtDNA mutation rates; these have centered on a date of about 200,000 years but with wide confidence intervals (Cann et al., 1987; Vigilant et al., 1991). The addition of a significant paternal contribution might well make this estimate considerably younger (Ross, 1991). The geographical patterning of mtDNA variation is consistent with an African origin, both from the great time depth of African lineages in Khoisan and Pygmy samples, and from the greater within-continent variation among the African lineages compared with those of any other region. This would again imply an African origin, given the assumption of relatively steady mutation rates for mtDNA change.

The calibration of nuclear DNA evolution is much more problematic than that of mtDNA, partly because of the greater number of genes to be considered and the more complex pattern of inheritance. A high diversity inside Africa is not found as consistently as in mtDNA analyses, and there are often very divergent non-African samples, such as New Guinea, or native Australians. Nevertheless, the basic pattern of relationships that often emerges is of an African **outgroup** followed by a division of the other populations into western Eurasian, east Asian/native American, southeast Asian, and Australian components. These results are based on single **locus** systems, such as beta globins (Wainscoat et al., 1986; Long et al., 1990) and on larger compilations of many different systems such as **blood groups** and other DNA **poly-**

morphisms (Cavalli-Sforza et al., 1988; Nei and Ota, 1991; Bowcock et al., 1991). The consistency of this pattern may provide clues as to the basic path followed during recent human evolution, although the detailed picture may be one of migration and gene flow, not at the origin of modern humans, but subsequently, among fully modern populations.

■

INTERPRETING THE DATA

So far, the majority of Pleistocene paleoanthropological data have accumulated from Europe and the Middle East, with a reasonable coverage for only parts of Africa, the Far East, and the Malay Archipelago. However, a find such as Jinniu Shan will have a major impact on our understanding of middle Pleistocene human evolution, whether or not the claimed age of about 250,000 years can be confirmed by further work. Although the partial skeleton remains unpublished in any detail, it is evident that the discovery does not fit easily into either an extreme multiregional or "Out of Africa" viewpoint. Its so-called advanced features contrast markedly with those of Chinese late *Homo erectus* from the contemporaneous or slightly older sites of Zhoukoudian and Hexian, and make a straightforward direct evolutionary relationship between them—as required by the multiregional model—difficult to contemplate (Stringer, 1990a). However, these same advanced features are also unexpected in China well before the appearance of modern humans in the region, as in the "Out of Africa" model. Either these features were spread by gene flow (from the west?) or they must have evolved locally. This would imply at least a dual origin for modern humans, in Africa and in East Asia. My own research on the related Dali and Maba fossil hominids (I have not yet been able to study Jinniu Shan) suggests that they are indeed more similar in cranial shape to modern humans than are the Neanderthals of Europe or the Middle East. Yet they are not as similar to modern humans as are approximately contemporaneous African specimens such as Jebel Irhoud and Ngaloba. A relatively close relationship between African and East Asian hominids of the later middle Pleistocene is certainly implied by these analyses, either through recent common ancestry or gene flow. Clearly, more data are required in order to establish the evolutionary significance of these Asian specimens. However, provisionally, I still consider the late archaic African sample to be the most plausible one for the ancestry of the first anatomically modern humans (Figure 4.11). Modern humans may have first appeared outside Africa in the Middle East, as documented by sites such as Jebel Qafzeh (Figure 4.12).

The Kebara discovery certainly reinforces some existing views about the Neanderthals—in particular, their robustness, especially in males. Compared with what is known of earlier African humans, the general skeletal robustness is a shared feature, but the physique of the Kebara man was very different from the inferred ancestral African pattern. The African pattern of a relatively tall, slim-hipped and long-legged physique dates as far back as 1.5 million years, as seen in the Kenyan juvenile skeleton of *Homo erectus* (Walker, Chapter 2). This physique was apparently also present in the material from the site of Broken Hill, Zambia (probably about 200,000 years old). In contrast, the Kebara individual had a wider, longer, and bulkier trunk, and other Neanderthal skeletons indicate shorter lower legs and a smaller mean stature than in the African specimens. The most probable explanation for these differences lies in climatic adaptation, because the African specimens have body proportions

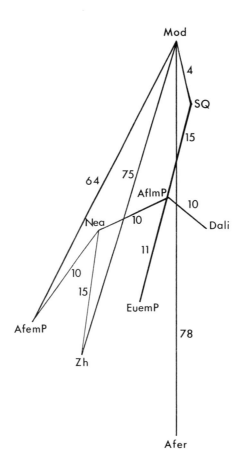

FIGURE 4.11

A phenogram (a diagram based on overall similarity) comparing various fossil skulls and a sample of recent humans. The numbers shown are distances obtained among the groups in a Penrose Shape analysis of 17 cranial measurements. In this kind of an analysis, the 17 different variables for each skull, or set of skulls, are mathematically combined to be represented by a single point. The distances between points correspond to the overall differences in shape as reflected by the 17 measures. Mod = means of a large and geographically diverse series of 2,540 recent crania measured by W. W. Howells; SQ = Skhul and Qafzeh (see Figure 4.9); Aflmp = African late middle Pleistocene specimens including Jebel Irhoud and Ngaloba (Figures 4.10, 4.12); Dali, as indicated (Figure 4.6); Nea = European late Neanderthals (Figure 4.7); EuemP = European early middle Pleistocene specimens such as Arago and Petralona; AfemP = African early middle Pleistocene specimens, such as Bodo and Broken Hill; Zh = late *Homo erectus* from Zhoukoudian, China; Afer = early African *Homo erectus* (Figures 2.7, 2.8). The Skhul and Qafzeh crania are, as expected, close to the modern mean sample. The nearest neighbor to both of these is the African late middle Pleistocene sample (AflmP), supporting an "Out of Africa" interpretation. However, the Dali specimen is also surprisingly close to the modern specimens.

like present-day Africans (approximating a slender cylinder, ideal for dissipating heat), while Kebara and other Neanderthals have physiques like present-day populations living in high latitudes or altitudes (that is, cold-adapted, approximating a spheroid, ideal for retaining heat) (Trinkaus, 1981; Ruff, 1991).

FIGURE 4.12

Cranium from Jebel Qafzeh viewed from the side. This specimen is dated to about 100,000 years ago and is a good example of the cranial appearance of the early African and Middle Eastern populations that may have been ancestral to all recent humans. Photo of a cast by the Photographic Unit, Natural History Museum, London.

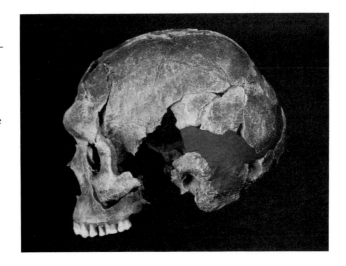

The Kebara skeleton also provides unique information about the form of the Neanderthal pelvis and allows an accurate reconstruction of pelvic shape and function for the first time (Rak, 1990). The wide pelvis has outwardly rotated **iliac blades**, indicating that the hip joints of the Kebara man did not operate in exactly the same manner as those of modern humans. Coupled with long-established differences in the shape of the femoral shaft, it appears that the Neanderthals walked differently from us, with more side-to-side tilt and perhaps less energetic efficiency. Given the new dating evidence from the Middle East, the fact that a more modern-looking pelvic structure already existed in the Qafzeh and Skhul individuals some 40,000 years before the Kebara man effectively excludes him and his Neanderthal kinfolk from an ancestral role for the first modern people. His robustness is not even suggestive of change in a modern direction through hybridization with contemporaneous modern individuals — one would have to look instead at the late Neanderthals of Amud in the Middle East, and at the Vindija Neanderthals in Europe for possible signs of this.

These late specimens can be combined with the archeological evidence of behavioral change in the last Neanderthals to suggest that there was some contact (and interbreeding) between late Neanderthals and intrusive early modern humans in Europe (in the "Out of Africa" model: Stringer and Grün, 1991; Stringer, 1992), or that a blending of local Neanderthal and extraneous African or Middle Eastern genes was in progress in eastern Europe (the gene-flow/assimilation model: Smith et al., 1989). However, the morphological and chronological evidence from Europe certainly does not support a straightforward multiregional interpretation because there seems to be a significant period of coexistence between late Neanderthals and early modern humans (Cro-Magnons) in Europe, with evidence that some late Neanderthals (such as the ones from Saint-Césaire) were showing little or no signs of either evolutionary or gene-flow-induced change in the direction of modern people (Figures 4.13, 4.14). The presence of supposed Neanderthal features in early Cro-Magnons has also been cited in support of an evolutionary relationship or gene flow between the respective populations (Wolpoff et al., 1984; Wolpoff, 1989; Smith et al., 1989). However, I find much of this evidence unconvincing. Indeed in some aspects of facial shape and body proportions early Cro-Magnons seem *less* like Neanderthals than do recent Europeans, perhaps indicating parallel climatic adaptations to that of Neanderthals *after* the Cro-Magnons appeared in Europe (Trinkaus, 1981; Stringer, 1989).

FIGURE 4.13

Comparison of a classic European Neanderthal (top) and an early anatomically modern human of Europe (bottom) viewed from the side. The Neanderthal specimen is the skull from La Chapelle-aux-Saints, France, found in 1908 and associated with Mousterian tools. It is of an adult male who suffered from arthritis and had lost most of his teeth in life, with very heavy wear on the few that were still retained. The anatomically modern cranium is from a rock shelter at Cro-Magnon, France, found in association with Aurignacian stone tools. Note the high forehead, slight brow ridges, more rounded brain case, and flatter face of the Cro-Magnon skull. Photo by C. Stringer, courtesy of the Musée de l'Homme, Paris.

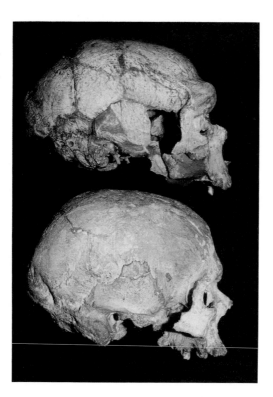

FIGURE 4.14

Comparison of Neanderthal (left) and Cro-Magnon (right) in anterior view. The Neanderthal specimen is La Ferrassie I. Photo by C. Stringer, courtesy of the Musée de l'Homme, Paris.

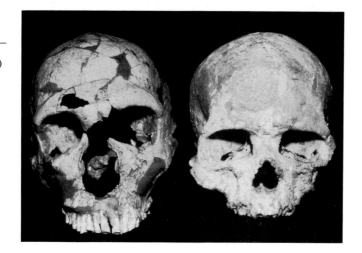

CONCLUDING REMARKS

Recent dating work reinforces the picture of modern humans appearing relatively earlier in Africa and the Middle East, and later in Europe, eastern Asia and Australia (later still, of course, in the Americas). Moreover, the most plausible ancestral samples for early modern humans, both in terms of morphology, and now also in estimated age, are from Africa (Grün and Stringer, 1991; Bräuer, 1992). However, we must remain cautious before fixing on Africa as *the* birthplace of modern humans from the fossil evidence alone. In the Middle East, early modern humans are present at about the same time as those of Africa, and we lack data on the local middle Pleistocene predecessors of the Skhul and Qafzeh people. The Tabun female appears, from electron spin resonance dating, to be approximately contemporaneous with, or somewhat older than, the Skhul and Qafzeh samples, but Tabun is definitely a Neanderthal, albeit primitive in certain respects (Grün et al., 1991). The Zuttiyeh craniofacial fragment is a more promising ancestral candidate, but it is too incomplete for detailed analyses. So we can neither confirm nor deny the possibility that the Middle East was also a zone of origin for modern humans, along with Africa. We must also remain cautious about the status of the Far East. So far there are no convincing claims for modern humans at early dates in this vast area, but the surprisingly advanced hominids from sites like Jinniu Shan and Dali are not far behind those of Africa in terms of their appropriateness as late archaic ancestors for modern humans. Full comparative analyses of these exciting finds are required first, followed by more fossil material from the period between 40,000 and 100,000 years ago. However, it is already possible to state from my current research that these later middle Pleistocene specimens do not seem to be uniquely linked to modern Oriental populations. This even remains true for the Zhoukoudian Upper Cave crania, about 25,000 years old. They seem unlike their modern regional successors and more closely resemble the contemporaneous Cro-Magnons of Europe.

Although I gradually became convinced of a recent African origin for modern humans through my research on the fossil record, strong support for this notion has come from geneticists, and, despite some recent doubts (for example, Templeton, 1993), I am sure that this trend will continue, with an increasingly dynamic and productive interaction between geneticists and paleontologists. I welcome this development because it will lead to an appreciation that the fossil and archaeological records are not redundant. No amount of genetic analysis of present-day populations could have revealed the existence of the Neanderthals in Europe as recently as 35,000 years ago, and data about their physique, pattern of evolution, and extinction is only available from the fossil and archaeological records. Even though the pace of new fossil discoveries will always be relatively slow and episodic, I hope I have shown that such discoveries can still raise new problems and suggest new solutions to old ones.

SUGGESTED READINGS

Cavalli-Sforza, L. L. 1991. Genes, peoples and languages. *Scientific American 265* (5): 72–78.

Klein, R. G. 1989. *The Human Career* (Chicago: Univ. of Chicago Press).

Putman, J. J. 1988. In search of modern humans. *National Geographic Magazine 174*: 438–477.

Reader, J. 1988. *Missing Links* (London: Penguin).

Stringer, C. B. 1990. The emergence of modern humans. *Scientific American 263* (6): 98–104.

REFERENCES

Aitken, M. J. 1990. *Science-based Dating in Archaeology* (London: Longman).

Bartstra, G. J., Soegondho, S., and van der Wijk, A. 1988. Ngandong man: Age and artifacts. *Journal of Human Evolution 17*: 325–337.

Bar-Yosef, O., Laville, H., Meignen, L., Tillier, A. M., Vandermeersch, B., Arensburg, B., Belfer-Cohen, A., Goldberg, P., Rak, Y., and Tchernov, E. 1988. Le sépulture néandertalienne de Kébara (unité XII). In: M. Otte (Ed.), *L'Homme de Neanderthal, Vol. 5: La Pensée* (Liége: ERAUL), pp. 17–24.

Beaumont, P. B., De Villiers, H., and Vogel, J. C. 1978. Modern man in sub-Saharan Africa prior to 49000 years BP: A review and evaluation with particular reference to Border Cave. *South African Journal of Science 74*: 409–419.

Bowcock, A. M., Kidd, J. R., Mountain, J. L., Herbert, J. M., Carotenuto, L., Kidd, K. K., and Cavalli-Sforza, L. L. 1991. Drift, admixture and selection in human evolution: A study with DNA polymorphisms. *Proceedings National Academy of Science, U.S.A. 88*: 839–843.

Bräuer, G. 1992. Africa's place in the evolution of *Homo sapiens*. In: G. Bräuer and F. H. Smith (Eds.), *Continuity or Replacement: Controversies in Homo sapiens Evolution* (Rotterdam: Balkema), pp. 83–98.

Bräuer, G., and Smith, F. H. (Eds.). 1992. *Continuity or Replacement: Controversies in Homo sapiens Evolution* (Rotterdam: Balkema).

Brown, M. 1990. *The Search for Eve* (New York: Harper and Row).

Butzer, K. 1976. The Mursi, Nkalabong and Kibish Formations, Lower Omo Basin, Ethiopia. In: Y. Coppens, F. C. Howell, G. L. Isaac, and R. E. Leakey (Eds.), *Earliest Man and Environments in the Lake Rudolf Basin* (Chicago: Univ. of Chicago Press), pp. 12–23.

Cann, R. L., Stoneking, M., and Wilson, A. C. 1987. Mitochondrial DNA and human evolution. *Nature 325*: 31–36.

Cavalli-Sforza, L. L., Piazza, A., Menozzi, P., and Mountain, J. 1988. Reconstruction of human evolution: Bringing together genetic, archaeological and linguistic data. *Proceedings of the National Academy of Science U.S.A. 85*: 6002–6006.

Chen, T., and Yinyun, Z. 1991. Palaeolithic chronology and possible coexistence of *Homo erectus* and *Homo sapiens* in China. *World Archaeology 23*: 147–152.

Day, M., and Stringer, C. B. 1991. Les restes crânien d'Omo-Kibish et leur classification à l'interieur du genre *Homo*. *L'Anthropologie 95*: 573–594.

Day, M., Twist, M., and Ward, S. 1991. Les vestiges post-crâniens d'Omo I (Kibish). *L'Anthropologie 95*: 595–610.

Deacon, H. J. 1989. Late Pleistocene palaeoecology and archaeology in the Southern Cape, South Africa. In: P. Mellars and C. B. Stringer (Eds.), *The Human Revolution: Behavioural and Biological Perspectives on the Origins of Modern Humans, Vol. I* (Edinburgh: Edinburgh Univ. Press), pp. 547–564.

Fagan, B. 1990. *The Journey from Eden: The Peopling of Our World* (London: Thames and Hudson).

Grün, R., Beaumont, P., and Stringer, C. B. 1990a. ESR dating evidence for early modern humans at Border Cave in South Africa. *Nature 344*: 537–539.

Grün, R., Shackleton, N. J., and Deacon, H. 1990b. Electron-spin-resonance dating of tooth enamel from Klasies River Mouth Cave. *Current Anthropology 31*: 427–432.

Grün, R., and Stringer, C. B. 1991. ESR dating and the evolution of modern humans. *Archaeometry 33*: 153–199.

Grün, R., Stringer, C. B., and Schwarcz, H. P. 1991. ERS dating of teeth from Garrod's Tabun cave collection. *Journal of Human Evolution 20*: 231–248.

Gyllensten, U., Wharton, D., Josefsson, A., and Wilson, A. C. 1991. Paternal inheritance of mitochondrial DNA in mice. *Nature 352*: 255–257.

Howells, W. W. 1976. Explaining modern man: Evolutionists versus migrationists. *Journal of Human Evolution 5*: 577–596.

Kennedy, K. A. R. 1979. The deep skull of Niah: An assessment of twenty years of speculation concerning its evolutionary significance. *Asian Perspectives 20*: 32–50.

Klein, R. G. 1989. *The Human Career* (Chicago: Univ. of Chicago Press).

Leakey, R. E. F., Butzer, K., and Day, M. H. 1969. Early *Homo sapiens* remains from the Omo River region of south-west Ethiopia. *Nature 222*: 1132–1138.

Lévêque, F., and Vandermeersch, B. 1981. Le Néandertalien de Saint-Césaire. *Recherche 12*: 242–244.

Long, J., Chakravarti, A., Boehm, C., Antonarakis, S., and Kazazian, H. 1990. Phylogeny of human beta-globin haplotypes and its implications for recent human evolution. *American Journal of Physical Anthropology 81*: 113–130.

Lu, Z. 1990. La decouverte de l'homme fossile de Jing-Niu-Shan. Première étude. *L'Anthropologie 94*: 899–902.

Mellars, P., and Stringer, C. B. 1989. *The Human Revolution: Behavioural and Biological Perspectives on the Origins of Modern Humans* (Edinburgh: Edinburgh Univ. Press).

Mercier, N., Valladas, H., Joron, J.-L., Reyss, J.-L., Lévêque, F., and Vandermeersch, B. 1991. Thermoluminescence dating of the late Neanderthal remains from Saint-Césaire. *Nature 351*: 737–739.

Nei, M., and Ota, T. 1991. Evolutionary relationships of human populations, at the molecular level. In: S. Osawa and T. Honjo (Eds.), *Evolution of Life: Fossils, Molecules and Culture* (Tokyo: Springer), pp. 415–428.

Pope, G. G. 1992. Craniofacial evidence for the origin of modern humans in China. *Yearbook of Physical Anthropology 35*: 243–298.

Rak, Y. 1990. On the differences between two pelvises of Mousterian context from the Qafzeh and Kebara caves, Israel. *American Journal of Physical Anthropology 81*: 323–332.

Rightmire, G. P., and Deacon, H. J. 1991. Comparative studies of Late Pleistocene human remains from Klasies River Mouth, South Africa. *Journal of Human Evolution 20*: 131–156.

Ross, P. E. 1991. Crossed lines. *Scientific American 265* (*4*): 17–18.

Ruff, C. B. 1991. Climate and body shape in hominid evolution. *Journal of Human Evolution 21*: 81–105.

Santa Luca, A. P. 1980. The Ngandong fossil hominids. *Yale University Publications in Anthropology 78*: 1–175.

Schwarcz, H. P., Grün, R., Vandermeersch, B., Bar-Yosef, O., Valladas, H., and Tchernov, E. 1988. ESR dates for the hominid burial site of Qafzeh in Israel. *Journal of Human Evolution 17*: 733–737.

Schwarcz, H. P., Buhay, W. M., Grün, R., Valladas, H., Tchernov, E., Bar-Yosef, O., and Vandermeersch, B. 1989. ESR dating of the Neanderthal site, Kebara Cave, Israel. *Journal of Archaeological Science 61*: 653–659.

Singer, R., and Wymer, J. (Eds.). 1982. *The Middle Stone Age at Klasies River Mouth in South Africa* (Chicago: Univ. of Chicago Press).

Smith, F. H., Falsetti, A. B., and Donnelly, S. M. 1989. Modern human origins. *Yearbook of Physical Anthropology 32*: 35–68.

Smith, F. H., and Trinkaus, E. 1991. Les origines de l'homme moderne en Europe centrale: Un cas de continuité. In: J. J. Hublin and A. M. Tillier (Eds.), *Aux Origines d'Homo sapiens. Nouvelle Encyclopédie Diderot* (Paris: Presses Universitaires de France), pp. 251–290.

Stringer, C. B. 1988. The dates of Eden. *Nature 331*: 565–566.

Stringer, C. B. 1989. Documenting the origin of modern humans, In: E. Trinkaus (Ed.), *The Emergence of Modern Humans* (Cambridge: Cambridge Univ. Press), pp. 67–96.

Stringer, C. B. 1990a. The Asian connection. *New Scientist 1743*: 33–37.

Stringer, C. B. 1990b. The emergence of modern humans. *Scientific American 263* (*6*): 98–104.

Stringer, C. B. 1992. Replacement, continuity, and the origin of *Homo sapiens*. In: G. Bräuer and F. H. Smith (Eds.), *Continuity or Replacement: Controversies in Homo sapiens Evolution* (Rotterdam: Balkema), pp. 9–24.

Stringer, C. B., and Andrews P. 1988. Genetic and fossil evidence for the origin of modern humans. *Science 239*: 1263–1268.

Stringer, C. B., and Grün, R. 1991. Time for the last Neanderthals. *Nature 351*: 701–702.

Stringer, C. B., Grün, R., Schwarcz, H. P., and Goldberg, P. 1989. ESR dates for the hominid burial site of Es Skhul in Israel. *Nature 338*: 756–758.

Templeton, A. R. 1993. The "Eve" Hypothesis: A genetic critique and reanalysis. *American Anthropologist 94*.

Trinkaus, E. 1981. Neanderthal limb proportions and cold adaptation. In: C. B. Stringer (Ed.), *Aspects of Human Evolution* (London: Taylor and Francis), pp. 187–224.

Trinkaus, E. 1989. (Ed.) *The Emergence of Modern Humans: Biocultural Adaptations in the Later Pleistocene* (Cambridge: Cambridge Univ. Press).

Valladas, H., Joron, J. L., Valladas, G., Arensburg, B., Bar-Yosef, O., Belfer-Cohen, A., Goldberg, P., Laville, H., Meignen, L., Rak, Y., Tchernov, E., Tillier, A. M., and Vandermeersch, B. 1987. Thermoluminescence dates for the Neanderthal burial site at Kebara in Israel. *Nature 330*: 159–160.

Valladas, H., Reys, J. L., Joron, J. L., Valladas, G., Bar-Yosef, O., and Vandermeersch, B. 1988. Thermoluminescence dating of Mousterian 'Proto-Cro-Magnon' remains from Israel and the origin of modern man. *Nature 331*: 614–616.

Vandermeersch, B. 1981. *Les Hommes fossiles de Qafzeh (Isräel)* (Paris: CNRS).

Vigilant, L., Stoneking, M., Harpending, H., Hawkes, K., and Wilson, A. C. 1991. African populations and the evolution of human mitochondrial DNA. *Science 253*: 1503–1507.

Wainscoat, J. S., Hill, A. V. S., Boyce, A. L., Flint, J., Hernandez, M., Thein, S. L., Old, J. M., Lynch, R. J., Falusi, A. G., Wetherall, D. J., and Clegg, J. B. 1986. Evolutionary relationships of human populations from an analysis of nuclear DNA polymorphisms. *Nature 319*: 491–493.

Wolpoff, M. H. 1989. Multiregional evolution: The fossil alternative to Eden. In: P. Mellars and C. B. Stringer (Eds.), *The Human Revolution: Behavioural and Biological Perspectives on the Origins of Modern Humans, Vol. I* (Edinburgh: Edinburgh Univ. Press), pp. 62–108.

Wolpoff, M. H., Smith, F. H., Malez, M., Radovcić, J., and Rukavina, D. 1981. Upper Pleistocene hominid remains from Vindija Cave, Croatia, Yugoslavia. *American Journal of Physical Anthropology 54*: 499–545.

Wolpoff, M. H., Wu, Z., and Thorne, A. G. 1984. Modern *Homo sapiens* origins: A general theory of hominid evolution involving the fossil evidence from east Asia. In: F. H. Smith and F. Spencer (Eds.), *The Origins of Modern Humans* (New York: Alan R. Liss), pp. 411–483.

CHAPTER 5

HUMANS AS MATERIALISTS AND SYMBOLISTS: IMAGE MAKING IN THE UPPER PALEOLITHIC

■

Margaret W. Conkey*

■

INTRODUCTION

We all tell stories about the past; this is the process of interpretation, the process of making inferences from archeological remains, using a wide variety of theories and methods. I don't use the phrase "telling stories" in a negative or antiscientific sense. All of science tells stories, or what Terrell (1990) calls "factual narratives" (Gergen and Gergen, 1986; Landau, 1984). Some people tell better stories than others, and the agreed-upon stories often change, such as those about the cave paintings, engravings, and other imagery that we call "Paleolithic art." Our stories about these images have sometimes changed because of new finds, new data, new facts; but they have also changed because of new assumptions about the makers, or because new interpreters have taken up the challenges of interpretation. Most stories about the art and imagery of late Ice Age Europe have focused on what the images mean (or might mean). In this chapter I twist the question a bit; instead of asking, What do they mean?, I want to ask, How were they meaningful?, To whom?, Why?, In what contexts?

Most modern viewers of Paleolithic art are convinced that this imagery must be symbolic, full of deep meanings, and perhaps even religious. But in wrestling with the questions about how the images were meaningful, I think we will see that the images we label Paleolithic art are just as striking and evocative and compelling to us because of their materiality and their materialness. To me, the simultaneous appropriation of materials for symbolic and social life and action renders these images of Paleolithic art so interesting and relevant. In one particular part of the globe (centered in Europe and extending eastward into Russia), when humans were nearly everywhere, this use of materials for symbolic and social life makes the imagery compelling.

Department of Anthropology, University of California, Berkeley, CA 94720

FIGURE 5.1

Polychrome wall panel in the cave of Lascaux in southwestern France showing superpositioned images of wild horse and bull. Reproduced courtesy of the National Prehistory Center, Périgueux, France.

FIGURE 5.2

Engraved image of an herbivore head, probably a chamois, surrounded by other engraved lines and shapes on a cave wall, Massat, Ariège, France. Photo by J. Clottes.

These peoples (or at least some of them) really understood their material worlds — properties and possibilities of pigments, bone, ivory and antler, clays, chemicals, liquids, stones, and rock types. The interplay between technologies (taken in the widest sense) and the symbolic, between technologies and ideas, makes this particular set of materials exciting anthropologically.

In reviewing the interpretation of these images, I stress that there are at least two aspects to interpretation. On the one hand, interpretation is a practice — we are the interpreters bringing specific ideas and questions to our interpretive enterprise (Rabinow and Sullivan, 1987) — and on the other hand, interpretation has results, in that there are interpretations to be made. There have been two monolithic and encompassing interpretations of the corpus of materials we call Paleolithic art, which have persisted into the 1980s. There is the idea of the art as part of sympathetic hunting magic, and the idea that the art is part of an underlying grammar or symbolic expression of world view. I note these two interpretations briefly in the first part of the chapter, primarily as a way to draw attention to three features that neither of these two monolithic interpretations came to terms with: (1) the diversity of the imagery itself, and of its possible contexts; (2) the idea that there may be no single interpretation to be had — especially for imagery that may have been developed and used for some 20,000 or more years (!); and (3) these images cannot be understood primarily as the origins of art, but only as part of particular and historical cultural products and practices.

In the second part of the chapter, I consider what happens if one takes these three observations into account, so that one comes to work with the very materials and their making. These are exciting times for the study of Paleolithic imagery precisely because of our break from past interpretations, because of the emergent recognition of the very diversity and richness of the images and materials themselves, and because researchers are actually more self-conscious, self-reflexive, and self-critical about the enterprise of interpreting these materials. I hope to make this clear even though demonstrating the richness and diversity of imagery would best be done using hundreds of slides and photos that cannot all be included in this book.

THE IMAGES AND THE INTERPRETATIONS

What Is the Subject Matter of Paleolithic Imagery?

Historically, we have used one label, Paleolithic art, to refer to a very large corpus of materials and images that span more than 25,000 years of time and that are found across a wide geographical area from the Iberian Peninsula to Russia. Although there may be a few images and forms that predate the **Upper Paleolithic** period, commencing about 35,000 years ago, there certainly seems to be a noticeable abundance of preserved images from this time through to the end of the last glacial period and the Upper Paleolithic, which ends about 10,000 years ago. This is something new in the archeological record, but exactly what this means about humans, their lives, and their activities is hotly debated (Conkey, 1983; Clark and Lindly, 1989; Mueller-Wille and Dickson, 1991; White, 1989). There are a lot of problems with a label like Paleolithic art. Such a global and inclusive term gives us no clue about geography, context, or even its temporal niche, which is only at the very end of the long, grand period of the Stone Age, or Paleolithic. And, of course, there are problems with calling all of the imagery "art."

Traditionally, analysts have talked about the imagery in two subcategories—wall art and portable art, each with fairly obvious referents. The portable art (or *art mobilier*) is assumed to have been potentially portable and movable, whereas the wall art is in place—fixed—on specific cave walls, so that makers and viewers must come to it. Recently, Vialou (1984) has proposed a third category—*art sur bloc*—to refer to those stone blocks, usually limestone, with engraving or sculpting that could, in theory, have been moved around within a locale. The idea is that these could have been semimobile and arranged or rearranged by various peoples. Still, the underlying metaphor for the categories is one of our making—a metaphor about mobility.

Above all, there has been a certain privileging of the painting: How often do you hear people equate Paleolithic art with cave painting? We even talk about the paintings being in different "galleries" (in the caves). It is no coincidence that the field of Paleolithic art research coincides with the late 19th and early 20th century sensibilities that we call modernism, which had, among other attributes, an obvious preference for and high value on painting. When you hear the name of the cave of Lascaux, you think of **polychrome** animals painted on the walls (Figure 5.1), but few people know that there are easily two to three times as many engravings on these cave walls. How many people know that there are at least four times as many images carved out of bone, antler, and ivory than there are painted on cave walls? And, as far as we can tell, the so-called portable art was probably much more a part of daily life than were the painted caves that have survived into the present, where there are few traces of occupation or living, and usually few traces of regular visits.

But no matter what categories we may devise for the imagery, the range of techniques used is indeed impressive, and it is surprising that so few scholars or thinkers before now have taken up the idea that the making, the technologies, might be a very crucial key for interpreting. A catalogue of the basic technologies would include engraving with stone tools, to form naturalistic (often animal) images or geometric forms (Figure 5.2). Many figurines were also made: some from fired earth (a kind of clay, as at Dolní Věstonice, Czechoslovakia, some 26,000 years ago), others

FIGURE 5.3

Carved figurine of a mammoth, from GeiSenklösterle, southwestern Germany. Reproduced courtesy of Dr. Joachim Hahn, University of Tubingen.

from ivory, antler, or even hard stone, such as **steatite** (Figure 5.3). Many different shapes and forms were perforated and these may have been suspended on sinew-type strings, sewn into clothing, or attached to other material objects. These perforated items include shells from both the Mediterranean and the Atlantic, the teeth of many animals, such as fox, and beads of ivory, bone, antler and stone (White, 1989).

Other kinds of engraving or carving include designs on stone as well as on bone and antler (Figure 5.4). At some sites, there are hundreds of stone plaquettes with many fine line engravings. Sometimes, cobbles or pebbles were engraved with precisely wrought animal forms. On some large limestone blocks or wall surfaces animals were sculpted in a bas-relief technique. Some images are nearly life-sized raised relief sculptures, such as at Angles sur l'Anglin or in the frieze of horses at Cap Blanc, Dordogne, France. And there are several striking examples of the use of clay to mold animal figures. The best known of these are the three little bison out of clay that were set up against an interior rock formation deep inside the cave of Le Tuc d'Audoubert in the French Pyrenees.

The other imagery on cave or rock shelter walls includes so-called painting, or the application of color, which was done in many ways. At Lascaux, for instance, some of the imagery was, more properly, drawn rather than painted, as the black outlines were created by the direct use of a charcoal stick (Figure 5.5). Some paintings appear to be a kind of flat wash technique, using liquids (often limestone-rich cave water) to blend the pigments into a workable medium (Figure 5.6 on page 100). Some pigments were even blown on, such as in some of the famous negative hand prints, where a bird-bone tube filled with pigments was used, or as Lorblanchet (1991) suggested, inspired by ongoing painting techniques of Australian aborigines, pigments may have been actually masticated directly in the mouth and blown on (Figure 5.7 on page 101). In some fascinating experimental work Lorblanchet has replicated the famous spotted horse image from the French site of Pech Merle using the technique of masticated and blown-on pigments. And in some images, painting and engraving are used together, such as on a large ox in Lascaux, where much of the animal outline is finely engraved.

There is a great variety in visual forms: some images of animals are complete, in-

FIGURE 5.4

Engraved designs on Magdalenian bone points. Reproduced courtesy of Randall White, New York University.

FIGURE 5.5

Polychrome image of wild bull at Lascaux, France. Reproduced courtesy of the National Prehistory Center, Périgueux, France.

cluding detailed treatments of fur and hooves. Other images are partial, only a head or torso. Most of the recognizable images are of animals, with such species as horse and bison the most frequent (Figure 5.8 on page 102, 5.9 on page 103), although there are also rhinos, fish, and even a weasel-like black outline figure (Figure 5.10 on page 103). There are many geometric shapes as well, for which we do not know the meaning or even the referent; these are all lumped under the category of "signs" (Figures 5.11, 5.12 on page 104). Several classification systems for the signs have developed, so that specialists talk about **tectiforms** or **claviforms**. The interpretations of the signs have, not surprisingly, varied according to the dominant interpretation of the art. For example, with the art-as-hunting-magic hypothesis, many geometric shapes were explained as related to hunting, such as traps, arrows, wounds, or hunters' huts!

There is quite a range of human depictions, although it has long been noted that the kind of realism (itself a very Western art historical term) or naturalism that characterizes many of the animal images is not characteristic of most human depictions. Exactly why humans tend to be depicted differently, as we see them, has never been satisfactorily explained. Perhaps it was the physical presence of humans (as artists, as viewers) that was enough for there to be a real humanness to the imagery; perhaps there was a different code for the depiction of humans; or perhaps there were (as yet not understood) metaphors for humans expressed in the animals and signs. However, humans are not entirely absent and, contrary to most popular Western texts or art books, there are many more kinds of human depictions than the famous so-called Venus figures (Figure 5.13 on page 105). There are recognizable images of males, but most human depictions can not readily or securely be assigned as male or female. Many are what we would call stick figures and others are only like humans, or anthropomorphs. There are some striking faces, such as in the famous caricatures from the engravings at La Marche (Pales, 1976; Hadingham, 1979), and yet many of the known figurines of females are notable because the face is not detailed in any way. As some of us argue, it may be in these features—what is or is not selected for depiction or emphasis—that we find some clues as to the various meanings or metaphors that may have inspired the imagery itself.

Our Western eyes are also often struck by some non-Western conventions, such as

FIGURE 5.6

Red horse lacking charcoal outline in the cave at Le Portel, Ariège, France. Photo by J. Clottes.

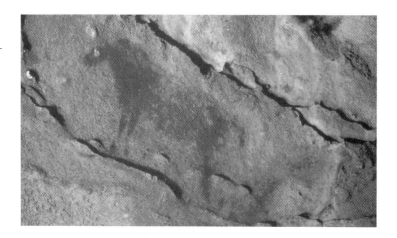

frequent superpositioning, where images appear to be piled one atop another in the same image-space. And there are rarely any of what we would call ground lines that, to our eyes, orient a figure and even place it in some sort of narrative space. Although there may be a dozen animal images, for example, in one general area of a cave (called a panel), these may be oriented at differing angles and it is not possible to say if this set of images could even be a scene (Figure 5.14 on page 106). There is really only one, and perhaps two, sets of images that could be interpreted to be a scene in the sense of illustrating some sort of an event or having different elements to a composition. This is the famous scene at the bottom of the well or shaft at Lascaux, where there appears to be a bison disemboweled by a spear, with an adjacent stick-figure male falling slightly backward, and an associated image of a bird on the top of a stick (Figure 5.15 on page 107). There are, however, as many interpretations of this supposed scene as there are interpreters!

Many of the images are very much involved with the shapes of the cave walls or the stone within which the images are made — and indeed this is a striking feature of much Paleolithic imagery (Figure 5.16 on page 107). One art historian has suggested that this shape-based nature or materiality of the images somehow works against the possibilities for the imagery to be narrative, or to be telling a story (Nodelman, 1985). He suggests that the images inhabit their material space so strongly that a fictive space is precluded. Yet, we should not, of course, expect that the images are necessarily literal and can be read directly; rather, if there is anything to be learned from the multitude of studies of art by anthropologists, it is that imagery can be about anything, and that much of it is more metaphoric than literal (Layton, 1992; Lewis-Williams, 1984).

Above all, despite this catalogue of images — of techniques, of basic forms, of conventions — note that these materials and images are not there all at once, as if they are necessarily part of some coherent, inclusive repertoire or package of image making. The very label, "Paleolithic art," has reinforced what I call the "collapse and composite" problem, where selected images and forms from different sites, different millennia even, are shown together, lumped together, and thus taken together as if they are necessarily associated in some meaningful cultural ways. These images are grouped on tenuous grounds, in that they span many, many generations and were surely used in widely varying contexts. Given that our maps lump art sites into one apparent tradition, it is no wonder how difficult it has been to break and enter into the block of prehistory we call the European Upper Paleolithic; but we must intervene into these constructed sets, these blocks of materials and implied

FIGURE 5.7

Handprint image formed by actually placing a hand on the cave wall and then blowing on pigment to create a negative image, El Castillo, Santander, Spain. Photo by M. Conkey.

cultural sources if we want to address the questions with which we began this chapter: How might the images have been meaningful, to whom, why, and in what contexts?

Distribution of Images in Time and Space

At the most simplistic level, there are differences in distributions, and these are begging for archeological explanations. Gamble (1982), for example, has drawn on what appears to be the delimited temporal, but widespread geographic, distribution of female figurines, centering around 26,000 years ago from France into European Russia. He has suggested that these might have been part of some symbolic system at work in establishing and reinforcing social alliance networks that he argues would have been a good way for people (as hunter-gatherers) to have helped ensure a reliable resource base under the particular environmental conditions of this time. Most of the cave painting that has been preserved seems to be attributed to the later millennia of the Upper Paleolithic, that is, to after 20,000 years ago. Furthermore, the painting seems to be particularly localized in southwestern Europe, especially in what is today France and Spain. Jochim (1983) has suggested that this cave painting period coincides with the onset of the last glaciation, when regions of present-day Germany would have been uninhabitable. Thus, it is not surprising, he suggests, that there is almost no evidence for painting on limestone cave walls in that region. Furthermore, he hypothesizes that the cave painting present in southwestern Europe may reflect more favorable areas for occupation — perhaps even refuge areas.

Of course, as archeologists we must always be careful about making too much from a presence or absence of archeological materials, especially given the always challenging factors effecting preservation. What may appear today to be well-defined distributions — such as cave painting often in deep uninhabited parts of caves — may have just as much to do with preservation (including the changing limestone topographies that could mean that today's cave entrances may not have been those of some 15,000 years ago) as with any real cave painting preferences. We are still being surprised by new discoveries that redefine what we thought about distributions. For example, there was the 1991 discovery of the Grotte Cosquer near present-day Marseille, which contains the first decorated walls so far east in southern France (*Life* magazine, November 1991). This cave is now accessible only from the Mediterranean Sea to a few extraordinarily skilled deep sea divers, but it would have been some distance inland from the sea during the last glaciation (about 18,000 years ago), when much of the oceans' water was taken up in the form of ice and glaciers, leading to lowered sea levels. Intriguingly, the one **carbon-14** date available so far for this cave, which comes from a hearth area on a bit of the cave surface not under water, is about 18,000 years. This does not, of course, necessarily date the paintings, but it does indicate human presence in the cave during the late Upper Paleolithic.

Another important broad difference in materials and images is that bone and antler working — not surprisingly — does not seem to have been uniform throughout the 25,000 years (or 1250 generations!) that we bracket off as the Upper Paleolithic. Rather, about 15,000 years ago, there does seem to be a kind of explosion of forms and imagery. There are elaborately decorated — even sculpted! — spear throwers and shaft-straighteners. There are little perforated cut-outs, or *decoupees*, depicting animal heads made from the distinctively shaped throat bone of the horse (Figure 5.17 on page 108). There are buttons or discs with images incised on them, which were

FIGURE 5.8

Outline drawing of a wild horse, Le Portel, Ariège, France. Photo by J. Clottes.

popped out of an animal shoulder blade. Although making spear points, rods, and other implements out of bone and antler is a hallmark of the Upper Paleolithic technologies from the earliest period, after 15,000 years ago these implements teem with what we would call decorations—incised geometric forms (chevrons, spirals, shell motifs) and figures of animals, anthropomorphs, and humans. In fact, many have noted that the range of different images, including even a grasshopper, on the bone and antler objects is far greater and more diverse than the repertoire of images on cave walls.

Without recounting further the details of the repertoires or the more refined temporal, spatial, and material differences (see Bahn and Vertut, 1988; Delporte, 1990; Sieveking, 1979), what is surprising, and what you should ask about, is why the interpretations for all of this imagery have so often been monolithic and inclusive. Given this array of media, of subject matter, of raw materials, of techniques, of differential distributions through time and space, what is surprising is that until the 1980s (and even into the present) the interpretations for Paleolithic art have been homogeneous and inclusive.

Images That Are "Good to Eat" or "Good to Think"?

Once the imagery was accepted officially in 1902 as being of genuine Ice Age antiquity—and there is a fascinating story about the reluctance of early prehistorians to believe these images were made by ancient people (Bahn and Vertut, 1988)—there was the development of the *foundation interpretation* (Conkey, in press). This baseline or foundation interpretation is the idea that the art and imagery was a manifestation of sympathetic hunting magic; that the animals and signs were created to ensure success in the hunt and of the fertility of the game (Figure 5.18 on page 109). Although this idea, promoted by the French researcher and priest, Abbe Breuil (for instance, 1952), has not completely gone away, the 1960s witnessed some serious challenges to many of its presuppositions about the imagery and image making. Influenced by the work of Annette Laming-Emperaire (1962) and Max Raphael (1945), Andre Leroi-Gourhan developed some ideas and compiled empirical documentation about the making and placing of images that revolutionized, in some senses, the study of not just Paleolithic imagery but the possibilities of archeological interpretation (see Leone, 1982; Conkey, 1989).

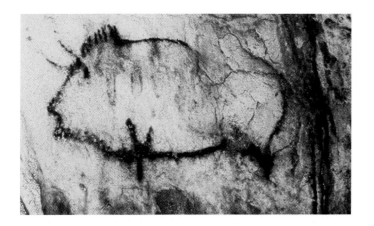

FIGURE 5.9

Drawing of a bison, Le Portel, Ariège, France. Reproduced from Sieveking, *The Cave Artists* (London: Thames and Hudson), 1979.

FIGURE 5.10

Outline drawing of an indeterminate animal that looks like a weasel or a stone-marten, Reseau Rene Clastres, Ariège, France. Photo by J. Clottes.

The 1960s work became known as the *mythogram interpretation* because, in its simplest form, these researchers suggested that the imagery was made and placed in certain locales because of some underlying mythological structures (about what the imagery meant or referred to) that were enacted through a "grammar" for image making. The imagery was shown to be dominated by certain species — what Leroi-Gourhan referred to as a "bestiary" because these select species were assumed to have certain relevant but unspecified symbolic properties. It was believed that there was such regularity in where images were placed in caves that one could speak of "an ideal sanctuary" (Leroi-Gourhan, 1965), and later, of a "formula" for the making and placing of images (Leroi-Gourhan, 1982; Arl. Leroi-Gourhan and Allain, 1979).

To summarize simply, we might say that the *hunting magic hypothesis* is a functionalist account; it suggests the function of the art in terms of some presumed basic needs (food). The idea is that animals are depicted because they were, to Upper Paleolithic peoples, a food source and were "good to eat." At the time, no one really documented or noted that there was not often a one-to-one correlation between the species depicted (mostly horse and bison) and those that dominated food refuse (often red deer or reindeer). Furthermore, the hunting magic hypothesis was developed when researchers and scholars were not yet comfortable with the idea that ancient hunter-gatherers could really be modern in certain senses; that is, these ancient peoples were thought to be prereligious and prescientific in the sense that they still somehow really believed in sympathetic wish-fulfillment activities and magic.

The 1960s ideas about the imagery came out of an intellectual movement called **structuralism**, which sought to define the underlying structures of human thought and action (see Lévi-Strauss, 1963; also Conkey, 1989, for a summary). This assumed that these peoples had fully modern human cognitive capacities and structures. The underlying structure, or mythogram, for Paleolithic art is premised on the idea that animals are "good to think," that is, they are good sources for metaphors. Rather than being "pre-us," the structuralist type of accounts assume that these Upper Paleolithic peoples were very much like us; they were, after all, modern in anatomy, being the same biological species as us, ***Homo sapiens sapiens***.

FIGURE 5.11

Geometric shapes or signs on cave wall, El Castillo, Spain. Photo by M. Conkey.

FIGURE 5.12

Geometric pattern of dots and lines, Marsoulas, France. Photo by J. Clottes.

Paleolithic Imagery Is Not the "Origins of Art"

One of the stories that underlies the shift in interpretation from hunting magic to mythogram is that scholars were changing their ideas about what difference among all humans is all about; *difference* was being renegotiated as was the relation of the past to the present. With the acceptance of the mythogram idea, the deep Upper Paleolithic past was brought much closer and under the rubric of "modern us". In the early 20th century, when hunting magic was eagerly endorsed as a way to account for the imagery, the hunter-gatherers of the late Ice Age were still very much distanced from the gentlemen scholars of Europe and were taken to be more savage than civilized. After all, the debate about the antiquity of Paleolithic imagery was mostly about the paintings—portable art objects had been more readily accepted as the products of ancient craftsmen—and painting was, in the late 19th and early 20th centuries indeed a hallmark of modernism and was not craft but art (Burgin, 1986).

Paleolithic imagery has long played a role—often a central one—in the debates about how and why anatomically modern humans (*Homo sapiens sapiens*) replaced Neanderthals in Europe (Chase and Dibble, 1987; Clark and Lindly, 1989; Conkey, 1983; White, 1989). It is therefore perhaps not so surprising that the imagery has been taken as part of a larger, almost utopian block of time—the Upper Paleolithic—when "Magdalenian hunters could kill an entire herd" (Pfeiffer, 1986; see Conkey, 1984). Human symbolic achievement—so-called art—had appeared complete with a "full-blown naturalism." All too often, the peoples of the European Upper Paleolithic have been construed to be an Eskimo-like group, frozen in their time-space tracks to stand for the origins of art and for some sort of explosion of symbolic behavior that is somehow associated with fully modern humans (Conkey, 1991).

FIGURE 5.13

A human figurine, or so-called Venus of Sireuil, Dordogne, France. Reproduced courtesy of the Musée des Antiquités Nationales, Saint-Germain en-Laye, near Paris.

Even though modern humans are now known to predate the European Upper Paleolithic by some 50,000 years or more (see Stringer's discussion in Chapter 4) the Upper Paleolithic—and all the modernity it implies—is still quite eurocentric. The 1986 cover of *Newsweek* proclaimed that the European Upper Paleolithic is "The Way We Were," with a subtitle of "Our Ice-Age Heritage: Language, Art, Fashion and the Family." The implication is that the categories, such as art, and social institutions of Western life and of anthropological analysis, such as the family, are already in place. The result of human evolution is clearly illustrated in the white Anglo-Saxon male whose image graces the cover of *Newsweek* (November 10, 1986). In contrast, the February 1992 cover of *Discover,* "The Magazine of Science," proclaims that Neanderthals had only "A Strange Life of Sex, Food, and Fireside Chats." Much about difference is being said here.

There is, of course, much to be said about the cultural assumptions and premises that still underlie the broad conceptualizations we hold of the human past, and many of these are clearly problematic in terms of fundamental issues such as sex and gender, race, class, and cultural or ethnic prowess. More precisely, there is much to be said about how interpretations of Paleolithic art have figured in these wider archeological narratives (see Conkey and Williams, 1991; Dobres, 1992). However, the last decade has certainly witnessed a break from more than a century's worth of work in Paleolithic imagery that has tended to interpret the materials in a monolithic way. For one thing, there is increasing recognition that there are many culturally standardized systems of visual representation (Munn, 1966). The animal head *decoupees*, for example, or the famous animal-ended spear-throwers out of antler, are striking examples of how there must have been one or a few individuals participating in a shared set of conventions about how to make certain images and forms. That 65% of all the imagery on cave walls is either horse or bison is not likely to be by chance. Thus, the overall figurative system that we have labeled Paleolithic art appears to be at least several, "perhaps interpenetrating," sign systems (Davis, 1986).

Thus, I suggest that we are at the point of a reconceptualization of all these cultural materials that we call Paleolithic art. These images are not just reflections of some broad cultural processes, such as symbolization. This is not necessarily art nor the origins of art. Not only is there increasing evidence for wall and portable images from elsewhere in the world at this same late Ice Age time (Bahn and Vertut, 1988), but this linking of the Eurorussian materials with *the* origins of art has had at least one major drawback for our interpretive possibilities. Paleolithic art has too long been taken as either the end point of human symbolic and cognitive evolution *or* as the beginning of human (that is, Western) art history—just open any introductory art history textbook and note the first chapter. This kind of positioning has left the Upper Paleolithic imagery and its contexts as a nonentity, as more of a part of some grand transformational scenario than as a set of rich and varied cultural materials embedded in their own particular historical contexts and meaning systems. In these grand schemes, what the images mean has only been sought at the macro level of being about the appearance of symbolism, of magic, or of cognitive modernity.

If, however, we accept that there are many sign systems, we cannot necessarily call it all art. This term presupposes the aesthetic and precludes investigation in terms that are less culturally loaded. The possibility of many sign systems demands, I believe, that we disarticulate the media and forms and images from a presumed cultural whole—"as art." We must try to understand the differing sign systems and, most important, to link the imagery to more specific social and historical contexts within which it was produced and reproduced.

FIGURE 5.14

Polychrome images of bison oriented at differing angles at Altamira Cave in Santander, Spain. Photo by E. Dominguez, reproduced courtesy of Cuevas de Altamira, Santander.

MATERIAL PRACTICES AND REPRESENTATION

If we want to link the images to more specific historical and social contexts, I would advocate working with the materials themselves. The images—in bone, antler, ivory, pigments—are superb archeological examples of the production and reproduction of material and symbolic repertoires. These materials were appropriated in various ways by people—men, women, and children—for varying symbolic and sociocultural reasons.

Although it is tempting to think of these materials as a way to support grand narratives about the appearance and so-called successes of modern humans, we must resist these uses and narratives. Clearly, anatomically modern humans came late to this geographic area, and this particular image making must first be explained in local, historical terms. As an archeologist, I would like to understand better why *these* materials, in *these* forms, occur in *these* sites, and at *these* varying times. The materials themselves seem to be particular sites where material practices and representation came together to form and re-form human experiences. And yet, for those archeologists who want to see the particular (these images) in terms of the more general processes of human evolution, the images and materials should elucidate how at least some late Pleistocene hunter-gatherers can be understood as *both* materialists and symbolists.

Technological Style and Pigment Processing

Several recent exciting studies of the images and their making offer a provocative baseline for our interpretive imaginations. For some years now, we have known that the mixing of pigments was a variable and sophisticated feature of some of the paintings, and in particular, those from the famous site of Lascaux (Arl. Leroi-Gourhan, 1982; Couraud and Leroi-Gourhan, 1979; Vandiver, 1983). The studies

FIGURE 5.15

A rare panel containing several images that appear to be part of a scene; from the cave at Lascaux, France. Reproduced courtesy of Hammond Inc., Maplewood, New Jersey.

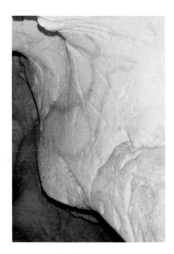

FIGURE 5.16

Use of natural shape of the cave wall to give body to an animal image, El Castillo, Santander, Spain. Photo by M. Conkey.

of the varied pigment preparation and uses show that some of the imagery at Lascaux is the result of a kind of a dialogue between the raw materials (pigments), the techniques for processing and application, and the images that were achieved. The paintings and drawings are there on the walls because of quite complex processing of pigments: grinding, mixing with cave water, mixing with what appears to be a powder of burned bone ash, for example. Many of what we might call the stylistic attributes of the images seem to be due, at least in part, to the ways in which the pigments were prepared and applied. Black, for example, often appears in the images as a long outline, the result of direct pigment application as in a drawing. Red, however, is often mixed into sort of a "slurry" and then used often for shading or filling in of imagery (Vandiver, 1983).

This intersection of techniques and images is what Lechtman (1977) would call "technological style." The very activities that generate or produce the artifacts and imagery are themselves patterned and stylistic. At the same time, the style of the resultant image (for example, a black-outline bull or a flat-wash red cow) is just as much a product of the technologies as of the symbolic and conceptual formations that also lie behind the making of the imagery. There is much to be learned about the technological style of Paleolithic material culture; or, as A. Leroi-Gourhan wrote in his 1982 book, "If plenty still remains to be discovered about the 'why?' of the symbols, almost everything remains to be said about the 'how?.' "

At Lascaux and at other sites, there is good evidence for reapplication of pigments, sometimes in different colors from the original. In some cases, there is an actual layer of recrystallized cave wall surface between pigment layers, suggesting that some time had passed since the original application. These pigment reapplications are in contrast to other sites where there seems to be good evidence for only one, quick visit. A most striking contrast to the polychrome, mixed-pigment imagery found at Lascaux (though not all imagery there was done in this way) is the inferences made by Lorblanchet (1980) about the making of the black outline figures that comprise the so-called Black Frieze in the French cave of Pech Merle (Figure 5.19 on page 109). After a careful study of these Black Frieze images,

FIGURE 5.17

Nineteen different specimens showing variation on a single theme, animal heads cut from horse hyoid (throat) bones from Labastide, France. Each head is about 5.5 cm in length, and they are all presumed to have formed the elements of a necklace. Photo by Robert Simmonet.

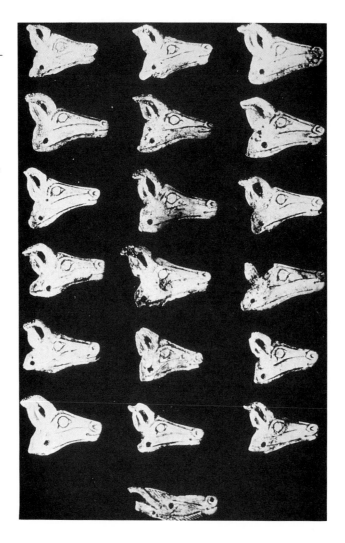

Lorblanchet carried out an experiment, in which he actually tried to replicate the image making of the frieze on an underground cave wall surface. He was able to infer from the Black Frieze that a set of geometric guidelines underlay each image and that the images were achieved by using the direct application of some manganese (black) to the cave wall surface. Rather than grinding and/or mixing the pigment to get a consistency that would stick to the walls, it appears that the image makers made a shallow trough with a stick into the wall surface that would then hold the pigment. Using a shallow stone hand-held lamp, much like the ones found in many caves, Lorblanchet replicated the figures according to the inferred guidelines, holding the manganese pigment in his other hand (Figure 5.20). To his amazement, he concluded that the figures could have been done in less than an hour. Although this is not definitive proof that the image makers did the figures so quickly, the process of image making here appears to have required relatively little time. Furthermore, the available archeological evidence, such as pollen preserved in cave sediments, suggested to Lorblanchet that the Black Frieze area was not visited frequently—in fact, probably a brief, one-time-only visit. This is quite different from the view we get of some caves, or from the earlier interpretations associated with the hunting-magic hypothesis; Pech Merle itself was once described by Breuil as a "sanctuary visited often by the faithful" (Lorblanchet, 1980). These contrasts, these differences between and within caves are particularly exciting and

FIGURE 5.18

Bison at the cave of Niaux, Ariège, France, with a straight line drawn across the torso, which has been interpreted as an arrow or spear. Drawings such as this one contributed to the interpretation that Paleolithic images served as hunting magic. Photo by J. Clottes.

provocative; they suggest varied uses, varied makers, varied purposes, and they demand that we try to understand what people were doing in these caves!

Pigment Recipes as Chronological and Behavioral Clues

Some of the most extensive work with pigments now is being carried out for sites and materials in the French Pyrenees (Clottes, in press). Similar pigment analyses are also being carried out in the Ardèche and Lot regions of France and in Cantabrian Spain. The Pyrenees is a fascinating geographic area for the study of Paleo-

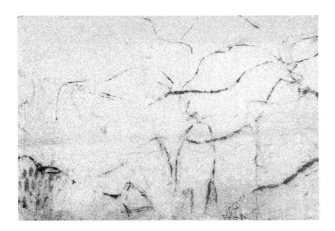

FIGURE 5.19

The Black Frieze in the cave of Pech Merle, Lot, France, depicting a mammoth (lower left) and other large animals and shapes. By permission of Anthony Cutler, Department of Art History, Pennsylvania State University.

FIGURE 5.20

Experiments at replicating Paleolithic techniques and the images themselves have been conducted in order to better understand the methods and the time involved in making the original images. Reproduced courtesy of M. Lorblanchet.

lithic imagery in its regional context because each site has a slightly different repertoire of media. Some have only wall art, another has only portable art and plaquettes, and yet there are some very distinctive, if not characteristic images and forms that are very Pyrenean, especially those from one of the two major periods of Upper Paleolithic occupation, the so-called Middle **Magdalenian** (about 14,000 years ago). My own research now focuses on what might be called the social geography of image making (Conkey, 1984), and these sophisticated studies of pigments and the distributions of different recipes for pigments is already one important line of evidence regarding the relationships and associations among and between sites. The pigment studies are one line of evidence about how these image-making peoples were using the region. As anyone who has tried to make archeological interpretations knows, we must use many lines of evidence to see how and where they converge toward reinforcing or challenging our interpretations.

The Pyrenean pigment work has certainly been enhanced by the relatively recent application of several analytical techniques now available to archeology. The study of materials is no longer a simple matter of identification based on surface features. Some of the techniques include **x-ray diffraction** and the use of the **scanning electron microscope** and **proton beam** to analyze the structure and elemental composition of pigment chemicals (Buisson et al., 1989; Clottes et al., 1990a, 1990b; Clottes, in press; Lorblanchet, 1990a, b). It is clear from the careful analysis of a variety of pigments that many of the paintings in the region (the Ariège, or eastern and mid-Pyrenees) are the result of sophisticated processing, but with enough replications or repeated procedures that we can detect some patterns and different recipes. These Magdalenian peoples deliberately added a range of materials as **binders**, so that the pigment would actually adhere to the cave walls: **potassium feldspar**, **biotite**, even **talc**. They also added **extenders**, so that the pigment would achieve a workable consistency. So far, most extenders appear to have been the use of cave water from underground puddles, lakes, or streams. Some, however, are of organic origin that make the pigments kind of oily; most organic sources can be identified so far as coming only from either plants (most of them) or animals (one instance, at the site of Fontanet).

The initial work was done at Niaux, famous for its Salon Noir paintings deep in the cave that has no evidence for human occupation (Clottes et al., 1990a, b). No hearths, food refuse, or **middens** of any kind exist within the cave, yet hundreds of painted images and signs occur there (Figures 5.21 through 5.24). Surprisingly, the analysts identified four different pigment recipes, some using **hematite** (red) but mostly relying on **manganese dioxide** and charcoal (black). In the Salon Noir itself, they found that very careful charcoal sketches underlay the final applications of pigments, and they found that the image makers used two slightly different versions of the same pigment recipe, suggesting two paint pots of the same basic mix.

The researchers have now applied these analytical techniques to paintings in other caves in the region, and there is great excitement about the questions to be pursued because these new compositional analyses need such small (if not downright tiny) amounts of the pigment that the paintings need not be destroyed. Thus, other labs and other tests can be applied to the same samples, and thereby confirm or extend the analyses based on minute fractions of pigment. Some questions thus spring to mind: Are paintings in the same cave made with the same pigments and recipes? Are there patterned differences in the making and using of pigments, such that horses, for example, are done differently than bison? Are all the examples of a given type of sign done with same pigment mix? Are what we think of as panels done with the same recipes? Is there integrity to the technologies of a panel? How do the pigment mixes and application methods in one cave relate to those in other nearby caves?

FIGURE 5.21

One of dozens of images that appear on the cave walls at Réseau Rene Clastres, Ariège, France, adjacent to Niaux. This one is of a bison subsequently overgrown by stalactites. This cave also contains the preserved footprints of Magdalenian-period people! Photo by J. Clottes.

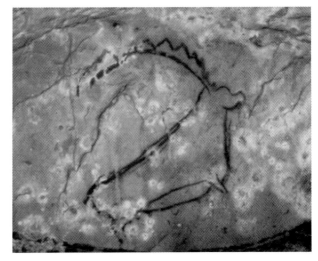

FIGURE 5.22

An ibex figure from Niaux Cave, France. Photo by J. Clottes.

Keeping in mind that previous scholars and theories assumed certain things about the painting processes that, in turn, were central to their interpretations, there is now some hope that we can move from assumptions to some intriguing empirical details. For example, Breuil thought that groups of figures in a panel and in a cave were gradually accumulated through time and by successive painters and even generations of painters. This assumption was important to Breuil's ideas about the imagery as hunting magic (the cave locale was sacred and each image was a new and separate magical act). Furthermore, this assumption was central to his

FIGURE 5.23

A Niaux horse. Photo by J. Clottes.

FIGURE 5.24

A Niaux bison. Photo by J. Clottes.

other major contribution: a stylistic chronology that placed the paintings into two successive temporal cycles of painting. The superpositioning of paintings, assumed to have been done one after another over stretches of time, was his key to the construction of the chronology (Breuil, 1952; Ucko and Rosenfeld, 1967). But the pigment analysis has already shown that, at least in some cases, a group of images were done with the same mix of the same recipe, suggesting that they are not individualized images accumulated over a longer time.

So far, in the French Pyrenees, there are studies under way for pigments in 12 different caves—on cave walls, portable art, and on blocks. As with the Black Frieze of Pech Merle, most of the Niaux and Reseau Clastres paintings seem to have been done quickly, and there is not much evidence for lingering or a drawn-out process of painting. At Lascaux, however, even scaffolding has been reported as part of the technology (Arl. Leroi-Gourhan, 1982). At Niaux alone, however, the varied pigment recipes and their suggested chronological significance (some have been associated with earlier Magdalenian painting and others with later painting) does disrupt and challenge the idea set in place by Leroi-Gourhan: namely, that the paintings are part of unified wholes done all at once or over a very short time. If the chronological attributions of different recipes are confirmed, this would give us the basis for understanding the painting process in a given locale. This also would test Leroi-Gourhan's idea that there is an "ideal sanctuary" and a rule-bound system for the placing of specific images within a cave.

Although the French researchers analyzing the Pyrenean pigment feel there is good evidence to support the idea that the variations in painting recipes are chronological markers, this does not necessarily rule out as well that other factors influenced the selection and use of certain pigment mixes. There might have been seasonal recipes or pigments associated with a certain event; or there might have been specific mixes associated with particular individuals, groups, genders, or clans. However, given that we still cannot date most Paleolithic art directly—although this is becoming a possibility—you may well wonder how we can infer chronological variations in pigment recipes.

Chronological marking is possible because of the correlation between the pigment recipes used for the cave wall images and those used on dated portable art objects, excavated in stratigraphic context. For example, portable images from La Vache (a cave located directly across the narrow valley from Niaux) include engraved objects with pigments deliberately rubbed into the engravings. There is other evidence at La Vache, such as the presence of grinding stones and pigment clumps, that attests to the fact that pigments were indeed processed here. Other datable pigment recipes come from pigments and objects found in datable stratigraphic contexts at the deep-cave habitation site of Enlene, which adjoins the famous decorated (but uninhabited) cave of Les Trois Freres. In this cave, most famous perhaps for the so-called sorcerer image that is part-human and part-animal, there are pigment mixes similar to those found at Enlene, confirming the suspected relationship between the habitation area of Enlene and the painting and engraving activities in Les Trois Freres.

From these sites of both the Middle and Late Magdalenian periods there seem to be recipes that characterize each chronological period. Of course, interpretations cannot be final, but only preliminary. Taking a bit of pigment from one part of a painting says nothing about the possibility that other parts of an image could have been done with other recipes, and certainly all images have not been sampled. But these are provocative beginnings. The preliminary clues about variation and chronology suggest that we can make detailed, empirical, and previously unimagined inferences about how these peoples conceptualized and used a wide variety of elements and features of their material world—the pigments themselves, and the minerals used as extenders and binders—that taken together were then further used to create images that most likely had cultural, metaphorical, individual, and collective meanings and implications.

This pigment work has teamed up, in a way, with another technical breakthrough, namely, the **accelerator-dating** method, that is giving us, for the first time, some direct and absolute dates for the paintings themselves. We have never before had absolute dates on any of the wall art; this has probably been a major contributing reason why rock art studies have not been taken seriously by scientific archeologists, for whom the establishment of a chronology is often the first order of business in any research. Until recently, we could date the wall art in a variety of ways, most of them based on what we call **relative dating** methods. These include associating a given style of wall art image with the same style found on portable objects retrieved from stratigraphic contexts and dated archeological levels. Other means included the ways in which some datable archeological deposits covered up cave walls bearing images; thus, the imagery had to predate the accumulation of the archeological deposits. Relative dating methods have been used to develop rather specific and widely adopted chronologies for the Paleolithic images. The two most accepted of these chronologies have been that of Breuil (the two cycles as mentioned above), and more recently, that of Leroi-Gourhan (1965, 1968) who identified four style periods.

The accelerator-dating method needs only minuscule amounts of organic material, such as from charcoal-based drawings, to provide an age estimate in actual numbers of years. At Niaux, for example, it is now known that one of the famous bison is dated to 12,440 ± 190 years ago (Clottes and Valladas, in press). The archeological deposits at the nearby site of La Vache, long thought to be at least somehow related to Niaux, date to 12,450 ± 105 years! The debate about some of the Niaux paintings as being earlier or later in the Magdalenian, on stylistic grounds, is in part resolved, in that at least one painting is confidently attributable to the later Magdalenian. But because Niaux has pigment recipes of both the early and later Magdalenian periods, it appears to have been a site that was revisited over the 3000+ years that we identify as the Pyrenean Magdalenian, and we must entertain the idea that the imagery here is not attributable to any *one* period, as has previously been advanced (Clottes, 1990, in press).

New Interpretive Possibilities

Although some people may view the pigment studies as merely the application of new scientific techniques to archeological data, I see these studies more broadly. They provide a baseline for realistically answering questions we have long asked; they are not merely new data or technical breakthroughs. They are, of course, nothing more than descriptive techniques when not considered in the context of questions about social action, human behavior, and cultural intentions. Many archeologists have wished to avoid asking about the intentions behind the image making. It all seems so slippery and impossible to prove. Yet, to understand intentions of a cultural product is "to interpret it as being in some sense oriented, structured to achieve certain effects" (Eagleton, 1983; Conkey, 1989). This, however, cannot be grasped apart from the practical conditions in which image making and image using operate. To understand intentions is to think about art (or whatever these images are) as a practice, rather than as an object, and practices do not exist without active individual human subjects.

The point is that researchers have now come to an interesting new interpretive position that could yield great insights into questions concerning the meaningfulness of these images and forms. Rather than focusing on the particular images (a horse, a tectiform) there is the possibility of gaining insights into the productive contexts within which—and somewhat more specifically, with the generative technologies by means of which—the imagery comes to be invested with at least some of its meanings. This involves a wider view of technology, one that takes technology as much more than a mediator between humans and resources, or as more than a mere vehicle for adaptation.

The ways in which materials are worked and the maintenance or changes in particular technological styles are often efficacious, nonverbal ways through which communities may enculturate, elaborate, and challenge all sorts of values and ideas (Lechtman, 1977, 1984). Among the interpretive possibilities before us, then, we have perhaps not appreciated the potential of the study of generative technologies. Obviously, there are many parameters involved in the creation of material culture: there are the structural properties and possibilities of the raw materials, and the cultural attitudes toward these; there is the social organization of labor (how many people to process pigments, to hold lamps, to erect a scaffold, and who, in the group, are they?); there are the technological modes of production, which may be additive, subtractive, carried out in many or in just a few steps; and there are the possibilities for ritual observances as part of the productive processes. Given such parameters, people do organize their technical behaviors along lines that are socially, economically, and ideologically meaningful. By not investigating generative technologies—how things are made, in the widest sense—we, as archeologists, would be ignoring one rich access route to interpreting past human lives.

■
CONCLUSIONS

I hope the next decade will witness a dramatic change in the study of Paleolithic art. For one thing, the label or term should become obsolete as we recognize the diversity and variety of the imagery and of its specific historical and social contexts. For many archeologists who chart the course—ever changing, it seems—of human evolution, these images and forms we call Paleolithic art have been both inspiring and elusive. We have been too ambitious perhaps in thinking we could account for it all as a symbolic explosion associated with modern humans; as hunting magic to ease the so-called precarious life of hunter-gatherers; as a long-standing, culturally enduring mythogram that both provided and reinforced certain cultural or even cosmological principles. These broad accounts may have been operative during all or part of the millennia over which the images were made and used, but other accounts may have been operative as well. These broad accounts thus have not helped us engage with the specific contexts and social acts that must have also been going on.

Undoubtedly, it is extremely difficult for archeologists to make inferences about what an artifact or an image meant to its creators and contemporaries. Perhaps we can never do it, especially in a so-called objective way distinct from the questions and concerns of the analysts in the present. Although I would never claim we can know with certainty *the* meanings of Paleolithic art, we can certainly make some interpretive moves toward an enriched and multifaceted understanding of what it might have meant. The recognition and empirical demonstration of the diversity of the images and their generative technologies should, however, be somewhat of a liberation from the confines of inclusive theories that try to capture *the* meaning of Paleolithic art. This brief descriptive chapter makes clear that this imagery is not all the same, and there is much more to the imagery and its making than polychrome animals and female Venus figurines. The image making is much more complex, more than the satisfaction of some aesthetic impulses of anatomically modern humans. It is more than a way of dealing with the sociocultural impacts of the last glaciation, more than the enactment of various sociopsychological influences, such as being impressed by the "dynamism of the hunt" or being "attracted to the female form," as some have described it.

Art, like language and many other so-called hallmarks of humanity, are unlikely to be all-or-nothing phenomena that suddenly (or gradually, depending on your evolutionary preferences) appear on the stage of human existence. Art often is a category from our own histories and experiences that renders the past comfortingly familiar. Although it is certainly justifiable for us to call these images art, in that they do often strike a resonant chord with what we think art is all about, we have to inquire into these images in other terms and from other perspectives if we are not merely to replicate the present. On the one hand, to most of us, the archeological enterprise of interpretation is supposed to find out more about the past, yet it is so often inextricably (and deceptively) really about the present. On the other hand, we only gain by recognizing the relation between the past and the present.

Thus, I would argue that these images of Paleolithic art are indeed what we might call sites where material practices and representation must have come together to form and re-form the varying human experiences of the people in the Paleolithic world of Eurorussia. Yet, because these are images that have (selectively) persisted into the 20th century and are a part of our experiences, dialogues, and cultural knowledge, these images are the same kinds of sites for us, where our material practices and our representations (of the past and of the present) come together to form and re-form our experiences. From the cultural products of nearly 30,000 years, both the materialness as well as the more heralded symbolic aspects of Paleolithic images and forms have captured our attention and fascination. In the detailing of the materialness, such as in the pigment processings, recent research has come to the brink of reengaging with the symbolic aspects that together now encourage new questions: How and to whom were the images meaningful? Why this imagery in those places, sites, forms and shapes, at those times? Inevitably, just trying to answer these questions will, in turn, generate greater insights into both the Upper Paleolithic worlds of Eurorussia, and into ourselves, which is, after all, what the search for knowledge is about.

SUGGESTED READINGS

Bahn, P., and Vertut, J. 1988. *Images of the Ice Age* (London: Bellew Publishing Co. Ltd.).

Conkey, M. In press. *Paleovisions: Interpreting the Imagery of Ice Age Europe* (New York: W. H. Freeman and Co.).

Leroi-Gourhan, A. 1982. *The Dawn of European Art* (Cambridge: Cambridge Univ. Press).

White, R. 1986. *Dark Caves, Bright Visions* (New York: American Museum of Natural History).

REFERENCES

Bahn, P., and Vertut, J. 1988. *Images of the Ice Age* (London: Bellew Publishing Co. Ltd.).

Breuil, H. 1952. *Four Hundred Centuries of Cave Art,* translated by Boyle, M. (Montignac: Centre d'Etudes et de Documentation Préhistoriques).

Buisson, D., Menu, M., Pinçon, G., and Walter, P. 1989. Les objets colorés du Paléolithique Supérieur: Cas de la Grotte de la Vache (Ariège). *Bulletin de la Société Préhistorique Française 86 (6)*: 183–191.

Burgin, V. 1986. *The End of Art Theory: Criticism and Post-Modernity* (Atlantic Highlands, NJ: Humanities Press International Inc.).

Chase, P., and Dibble, H. 1987. Middle Palaeolithic symbolism: A review of current evidence and interpretations. *Journal of Anthropological Archaeology 6*: 263–296.

Clark, G. A., and Lindly, J. M. 1989. Modern human origins in the Levant and western Asia: The fossil and archeological evidence. *American Anthropologist 91*: 962–985.

Clottes, J. 1990. The parietal art of the Late Magdalenian, translated by Bahn, P. *Antiquity 64*: 527–648.

Clottes, J. In press. Paint analyses from several Magdalenian caves in the Ariège region of France. *Journal of Archaeological Science.*

Clottes, J., and Valladas, H. In press. Datation des bisons de Niaux. *Bulletin de la Société Préhistorique Française.*

Clottes, J., Menu, M., and Walter, P. 1990a. La préparation des peintures Magdaléniennes des cavernes Ariègeoises. *Bulletin de la Société Préhistorique Française 87 (6)*: 170–192.

Clottes, J., Menu, M., and Walter, P. 1990b. New light on the Niaux paintings. *Rock Art Research 7 (1)*: 21–26.

Conkey, M. 1983. On the origins of Palaeolithic art: A review and some critical thoughts. In: E. Trinkaus (Ed.), *The Mousterian Legacy: Human Biocultural Change in the Upper Pleistocene*, BAR International Series No. 164: 201–227.

Conkey, M. 1984. To find ourselves: Art and social geography of prehistoric hunter gatherers. In: C. Schrire (Ed.), *Past and Present in Hunter Gatherer Studies* (Orlando: Academic Press), pp. 253–276.

Conkey, M. 1989. The structural analysis of Palaeolithic art. In: C. C. Lamberg-Karolovsky (Ed.), *Archaeological Thought in America* (Cambridge: Cambridge Univ. Press), pp. 135–154.

Conkey, M. 1991. Contexts of action, contexts of power: Material culture and gender in the Magdalenian. In: J. Gero and M. Conkey (Eds.), *Engendering Archaeology: Women in Prehistory* (Oxford: Blackwell), pp. 57–92.

Conkey, M. In press. *Paleovisions: Interpreting the Imagery of Ice Age Europe* (New York: W. H. Freeman and Co.).

Conkey, M., and Williams, S. 1991. Original narratives: The political economy of gender in archaeology. In: M. di Leonardo (Ed.), *Gender at the Crossroads of Knowledge: Feminist Anthropology in the Post-Modern Era* (Berkeley: Univ. of California Press), pp. 102–139.

Couraud, C., and Leroi-Gourhan, Arl. 1979. Les colorants. In: Arl. Leroi-Gourhan and J. Allain (Eds.), *Lascaux Inconnu* (Paris: CNRS), pp. 153–170.

Davis, W. 1986. The origins of image-making. *Current Anthropology 27*: 193–215.

Delporte, H. 1990. *L'Image des Animaux dans l'Art Préhistorique* (Paris: Picard).

Dobres, M-A. 1992. Re-presentations of Palaeolithic visual imagery: Simulacra and their alternatives. *Kroeber Anthropological Society Papers 73–74*: 1–25. Department of Anthropology, Univ. of California, Berkeley.

Eagleton, T. 1983. *Literary Theory: An Introduction* (Minneapolis: Univ. of Minnesota Press).

Gamble, C. 1982. Interaction and alliance in Palaeolithic society. *Man (NS) 17 (1)*: 92–107.

Gergen, K. J., and Gergen, M. M. 1986. Narrative form and the construction of psychological science. In: T. R. Garbin (Ed.), *Narrative Psychology* (New York: Praeger), pp. 22–44.

Hadingham, E. 1979. *Secrets of the Ice Age: A Reappraisal of Prehistoric Man* (New York: Walker and Co.).

Jochim, M. 1983. Palaeolithic cave art in ecological perspective. In: G. Bailey (Ed.), *Hunter-*

Gatherer Economy in Prehistory: A European Perspective (Cambridge: Cambridge Univ. Press), pp. 212–219.

Laming-Emperaire, A. 1962. *La Signification de l'Art Rupestre Paléolithique* (Paris: Picard).

Landau, M. 1984. Human evolution as narrative. *American Scientist 72*: 262–268.

Layton, R. 1992. *The Anthropology of Art*, 2d ed. (Cambridge: Cambridge Univ. Press).

Lechtman, H. 1977. Style in technology: Some early thoughts. In: H. Lechtman and R. Merrill (Eds.), *Material Culture: Styles, Organization, and Dynamics of Technology* (St. Paul, MN: American Ethnological Society), pp. 3–20.

Lechtman, H. 1984. Andean value systems and the development of prehistoric metallurgy. *Technology and Culture 15 (1)*: 1–36.

Leone, M. 1982. Some opinions about recovering mind. *American Anthropologist 47*: 742–760.

Leroi-Gourhan, A. 1965. *Treasures of Prehistoric Art* (New York: Harry Abrams).

Leroi-Gourhan, A. 1968. The evolution of Palaeolithic art. *Scientific American 209 (2)*: 58–74.

Leroi-Gourhan, A. 1982. *The Dawn of European Art* (Cambridge: Cambridge Univ. Press).

Leroi-Gourhan, Arl. 1982. The archaeology of Lascaux Cave. *Scientific American 246 (6)*: 104–113.

Leroi-Gourhan, Arl., and Allain, J. (Eds.) 1979. *Lascaux Inconnu* (Paris: CNRS).

Lévi-Strauss, C. 1963. *Structural Anthropology*, translated by Jacobson, C., and Grundfest Schoepf, B. (New York: Basic Books).

Lewis-Williams, J. D. 1984. The rock art workshop: Narrative or metaphor? In: M. Hall, G. Avery, M. L. Wilson, and A. J. B. Humphreys (Eds.), *Frontiers: Southern African Archaeology Today. Cambridge Monographs in African Archaeology 10*, BAR International Series No. 207, pp. 323–327.

Lorblanchet, M. 1980. Peindre sur les parois de grottes. *Dossiers de l'Archéologie 46*: 33–39.

Lorblanchet, M. 1990a. Étude des pigments de grottes ornées Paléolithiques du Quercy. *Bulletin des Soc. et Litt. Sci. Art du Lot CXI (2)*: 93–143.

Lorblanchet, M. 1990b. Palaeolithic pigments in the Quercy, France. *Rock Art Research 7*: 4–20.

Lorblanchet, M. 1991. Spitting images: Replicating the spotted horses of Pech Merle. *Archaeology 44* (Nov./Dec.): 24–31.

Mueller-Wille, C., and Dickson, D. B. 1991. An examination of some models of late Pleistocene society in southwestern Europe. In: G. A. Clark (Ed.), *Perspectives on the Past: Theoretical Biases in Mediterranean Hunter-Gatherer Research* (Philadelphia: Univ. of Pennsylvania), pp. 25–55.

Munn, N. 1966. Visual categories: An approach to the study of representational systems. *American Anthropologist 68*: 936–950.

Nodelman, S. 1985. Untitled. Paper presented at the College Art Association Annual Meetings, Los Angeles.

Pales, L. 1976. *Les Gravures de la Marche, II. Les Humains* (Paris: Ophyrus).

Pfeiffer, J. 1986. Cro-Magnon hunters were really us, working out strategies for survival. *Smithsonian 17 (7)*: 74–85.

Rabinow, P., and Sullivan, W. 1987. The interpretive turn: A second look. In: P. Rabinow and W. Sullivan (Eds.), *Interpretive Social Science: A Second Look* (Berkeley: Univ. of California Press), pp. 1–30.

Raphael, M. 1945. *Prehistoric Cave Paintings*, translated by Guterman, N., The Bollingen Series IV, Pantheon Books (New York: Marchbanks Press).

Sieveking, A. 1979. *The Cave Artists* (London: Thames and Hudson).

Terrell, J. 1990. Storytelling and prehistory. *Archaeological Method and Theory 2*: 1–29.

Ucko, P., and Rosenfeld, A. 1967. *Palaeolithic Cave Art* (New York: McGraw-Hill).

Ucko, P. J. and Rosenfeld, A. 1967. *Prehistoric Art* (London: Thames and Hudson).

Vandiver, P. 1983. *Palaeolithic Pigments and Processing.* Unpublished Master's Thesis, Department of Material Sciences, Massachusetts Institute of Technology, Cambridge, MA.

Vialou, D. 1984. Des blocs sculptés et gravés. *Histoire et Archéologie 87*: 70–72.

White, R. 1989. Visual thinking in the Ice Age. *Scientific American 261* (July): 92–99.

CHAPTER 6

CULTURE AND HUMAN EVOLUTION

■

Robert Boyd*

Peter J. Richerson**

■

INTRODUCTION

Unlike other species, humans rely heavily on culture as a means of adaptation. Like all other organisms we adapt genetically, and like most other animals, we adapt using individual learning. However, we also acquire a great deal of adaptive information from other conspecifics by imitation, teaching, and other forms of cultural transmission. In contrast, social learning is rare in other animals and limited to only a few characters in those species that have it (Zentall and Galef, 1988).

It is important to distinguish between culture and individual learning. Culture is often lumped with ordinary individual learning and other environmental effects under the heading of *nurture*, to be contrasted with genes—*nature*. This way of thinking is responsible for much confusion about the evolution of human behavior. Culture differs from individual learning because ways of behaving are acquired from other individuals. For the most part, humans do not learn their language, occupational skills, or forms of social behavior on their own, they learn them from parents, teachers, peers, and others. Cultural behaviors are more like genetic traits than are ordinary learned variations in behavior. Like genes, they are inherited and transmitted in a potentially endless chain, but in the absence of culture, behaviors acquired by individual learning are lost with the death of the learner.

Because culture is transmitted, it can be studied using the same Darwinian methods used to study genetic evolution. Human populations transmit a pool of culturally determined behavioral variation that is cumulatively modified to produce evolutionary change, much as they transmit an evolving gene pool. To understand cultural change we must keep track of all the processes in the lives of individuals that

*Department of Anthropology, University of California, Los Angeles, CA 90024

**Institute of Ecology, University of California, Davis, CA 95616

increase the frequency of some **cultural variants** and decrease the frequency of others. Of course, these processes do differ substantially from the processes of genetic evolution. Most important perhaps, culture allows inheritance of behavior acquired during a lifetime; individually learned variants will not be lost with death, but rather, can be taught or imitated.

Culture can also be studied using the substantive conclusions of Darwinism, that the process of selection will lead to adaptation. It is plausible that natural selection has shaped the human psyche so that people tend to acquire adaptive beliefs and values, however it is that culture may work in detail. If not, how can we account for the evolution of the complex, costly organ that manipulates culture, the human brain? To the extent that this premise is correct, human behavior can be predicted using theory drawn directly from **behavioral ecology**, and no special account need be taken of the processes by which people acquire that behavior. The gambit of ignoring the details of how genes, learning, and other factors actually produce adaptive behavior has proven to be very successful in the study of the behavior of other animals. The substantive use of Darwinism to understand behavior has also been used, with some success, by human **sociobiologists** (see Smith and Winterhalder, 1992, for a review of this work).

We believe that many important aspects of human evolution can only be understood by combining both the methods and substance of Darwinism. There is little doubt that the organic capacities underlying human learning and behavior were shaped by natural selection and, thus, that the behavior resulting from these capacities must have been adaptive, at least in past environments. Nonetheless, attention to the processes of cultural evolution is important for understanding human evolution. First, the rate at which a population adapts to changing circumstances depends on the mechanism of adaptation. Genetic adaptation by natural selection is a relatively slow process, individual learning is fast, and as we shall see, cultural adaptation may range from one extreme to the other. For some disciplines, such as archeology, the rate of adaptation may be of great interest.

Second, we argue at length that cultural adaptation can yield qualitatively different outcomes from those predicted by conventional **fitness optimizing theory**, even if one assumes that the capacity for culture has been shaped solely by natural selection acting on genetic variation. An evolutionary explanation of the large scale of human cooperation, for example, may require an understanding of how culture alters the evolutionary process. In this insistence on the importance of applying the methodological tools of Darwinism to cultural evolution we part company with many human sociobiologists, although it seems to us that this requirement is only a necessary and friendly amendment to the substantive use of Darwinism.

CULTURAL EVOLUTION IS DARWINIAN

Life Cycles and Recursion Models

How are Darwinian methods used to understand cultural evolution? The first step in building a Darwinian model of evolution is to specify the life cycle for the organism under consideration. For some trait that we might wish to study the life cycle shown in Figure 6.1 might suffice. During the course of a single generation, various events occur in the lives of individuals in the population. Genetic transmission creates **zy-**

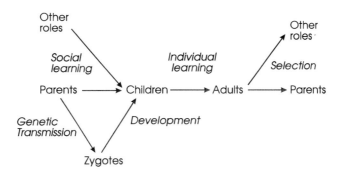

FIGURE 6.1

A diagram of the life cycle described in the text. Individuals acquire genetic variants from their parents, and cultural variants from their parents and individuals occupying other social roles. During maturation individuals may learn new cultural variants or modify existing ones. Finally, individuals characterized by some cultural variants may be more likely than others to attain social roles necessary for genetic or cultural transmission.

gotes that develop into children who are **encultured** by a set of individuals who occupy certain social roles. For example, for many cultural traits the biological parents may play an important role, while for other traits individuals occupying other roles such as grandmother, teacher, or priest may be important in enculturation. Children learn new beliefs and values or modify old ones as they mature to become adults. Then these adults live their lives, interacting with the physical and social environment. Some of these adults become parents and attain other social roles involved in cultural transmission. The next step is to describe what happens to individuals during each step in the life cycle, and how these events cause the frequency of some cultural variants to increase and others to decrease. Finally, by combining the effect of each step in the life cycle, we obtain a model of how culture changes over one generation.

A virtue of this picture of the cultural evolutionary process is that it is easily formalized as a mathematical model. An equation derived from the verbal model described in the last paragraph is called a **recursion.** Recursions capture the nature of the evolutionary process on a microscopic time scale. The objective, however, is to predict the long-term course of evolution. To accomplish this the recursion is iterated, stepping forward from generation to generation to determine how the frequency of different cultural variants changes over the long term. Mathematical models of this form have played an important role in the study of biological evolution because the concatenation of evolutionary processes affecting many individuals over a long period of time tends to escape easy analysis by unaided intuition. Mathematical models are a useful mental prosthesis to help us think more clearly about the rather intricate logic of evolutionary mechanisms. Much of what follows here is shaped by our interpretations of the mathematical models of cultural evolution that we and our colleagues have constructed.

We call the distinct processes that cause the frequency of cultural variants to change *forces.* In the remainder of this section we consider two such forces, **guided variation** and **biased transmission**. These forces are important because they may often cause culture to be adaptive in the substantive Darwinian sense.

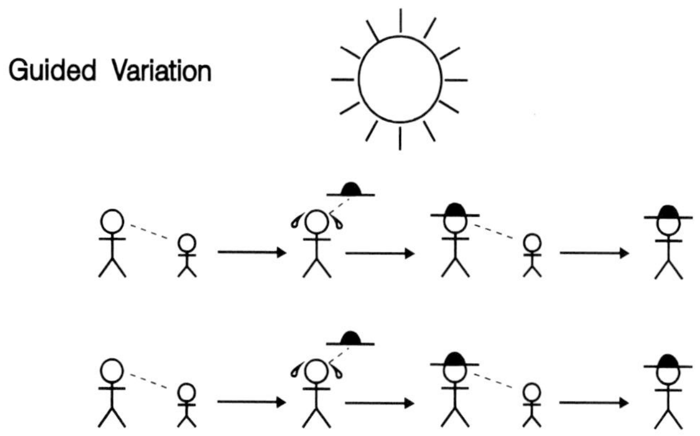

Guided Variation

FIGURE 6.2

A schematic representation of guided variation. Assume that there are two cultural variants: wearing a hat and going bare headed. Young people initially acquire their sartorial habits by copying the behavior of a single model. In the particular population pictured, adults do not wear hats. As they mature, individuals learn that in the current sunny environment hats are desirable, and they adopt the habit of wearing hats. During the next generation young individuals copy this new behavior.

Learning

We begin with the effect of individual learning. Because culture is acquired by copying the **phenotypes** of other individuals, culture allows the inheritance of new variation acquired during a lifetime. Individuals acquire beliefs and values by social learning. Such culturally acquired information is often affected by individual learning during the individual's life. People may modify existing beliefs, or even adopt completely new ones, as a result of their experiences. When such people are subsequently imitated, they transmit the modified beliefs. The next generation can engage in more individual learning and change the trait even further. When the beliefs of one generation are linked to the next by cultural transmission, learning can lead to cumulative, often adaptive, change. We say that such change results from the force of *guided variation* (Figure 6.2).

If individual learning is not to be random, there must be some rules that govern which behaviors are acquired and which are rejected. The strength and direction of guided variation depend on the nature of these learning rules. **Operant conditioning** provides a good example of how such rules work. An animal's nervous system causes some environmental events to be reinforcing and others to be aversive. The behavioral variation that individuals exhibit is shaped by such stimuli so that reinforced (generally adaptive) behaviors are retained, while those that result in aversive stimuli (normally maladaptive) disappear. Other forms of individual learning involve more complex, cognitively mediated rules. In every case, however, the kinds of traits acquired by learning depend on rules expressed in the nervous system that were acquired genetically or during an earlier episode of cultural transmission.

Thus, the kinds of traits increased by guided variation depend on the nature of the evolutionary forces that shaped the learning rules. The case in which learning rules are genetically transmitted and shaped by natural selection is of particular interest. First, this is the primitive case and thus is important for understanding the evolutionary origin of guided variation. Second, evidence suggests that such rules remain important influences on human behavior (Lumsden and Wilson, 1981; Tooby and

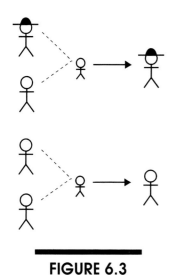

FIGURE 6.3

A schematic representation of directly biased transmission. Again there are two cultural variants: wearing a hat and going bare headed. Young people acquire their sartorial beliefs by observing the behavior of two models. They do not imitate blindly, however; they try out each behavior they observe and retain the one that seems best. Thus, if a young individual is exposed to one model that wears a hat and another that does not, he chooses the habit of wearing a hat because (by assumption) it is better in the local environment. However, if both models do not wear hats, then individuals acquire the habit of going bare headed because there is nothing to compare. As a result the strength of biased transmission depends on the frequency of alternative behaviors in the population. If most individuals are the same, then the composition of the population changes slowly because most sets of models have the same behavior. The frequency changes more rapidly when both behaviors are common. It also changes faster when sets of models are larger so that rare favorable variants occur in more sets.

Cosmides, 1989; Cosmides and Tooby, 1989). Finally, if learning rules were shaped by guided variation or some other force of cultural evolution, we then must ask how those prior learning rules were acquired. A chain of cultural rules followed backward through time will often end in genetically acquired traits of some kind.

Adaptation

People do not just imitate others at random and then modify behavior on the basis of their own experience; they also pick and choose whom and what to imitate in the first place. We call this process *biased transmission*. The simplest form of biased transmission, **direct bias**, can make use of the same guiding motivations as are used in guided variation. But with direct bias, there is no need to invent or reinvent the behavior concerned, but only to evaluate alternative behaviors and choose among them (Figure 6.3). If behavior is at all complex, it is much easier to evaluate available alternatives than it is to invent for oneself. Plagiarism is usually easier than invention, so the distinction between these two forces is not trivial. We consider another form of biased transmission subsequently.

Guided variation and biased transmission allow populations to adapt relatively quickly and effectively to changing environments. This is easiest to see when the goals of the learning rules are closely correlated with genetic **fitness**. If human foraging practices are adopted or rejected according to their energy payoff per unit time, then the foraging practices used in the population will adapt to changing environments almost as if natural selection were responsible. If the adoption of foraging practices is strongly affected by consideration of prestige, say that associated with male success in hunting dangerous prey, then the resulting pattern of behavior may be different. However, there will still be a pattern of adaptation to different environments, but now in the sense of increasing prestige rather than calories.

The rate at which a population can adapt by guided variation or biased transmission depends on how hard it is to evaluate alternative behaviors. If a new crop variety has substantially higher yields than older cultigens, then it is easy for farmers to detect the difference. However, the benefits of many other desirable traits may be hard to detect. The practice of boiling drinking water substantially reduces infant mortality due to diarrhea. Nonetheless, the practice may fail to spread because its effects are confounded by many other sources of diarrhea, because it conflicts with folk medical theory, and because the microbial causative agents killed by boiling are invisible. It may often be difficult to determine which variant is best, even if different variants have very different fitnesses. Traits whose net beneficial effects are only apparent when averaged over substantial periods of time may be especially difficult to evaluate. Even when learning is difficult, however, culture can accumulate small learned steps over many generations, leading to larger changes than would be possible when each generation has to learn anew.

The Diffusion of Innovations

Studies of the diffusion of technical innovations illustrate how the strength of guided variation and direct bias vary in response to circumstances. It is well known that humans make extensive use of pragmatic decision-making techniques when considering adopting potentially useful innovations. Rogers (1983) reviewed many studies suggesting that the perceived advantage of new technology relative to old is one of the most important variables in explaining why particular innovations spread. In conformance with the theory, people with more education and more resources are more likely to be the early adopters of innovations. The difficulty of evaluating innovations, and the impact of costly errors, weigh more heavily against less educated and poorer people adopting innovations on the basis of their own evaluations. They, sensibly, wait for those who can better bear the costs of independent decision making to try them out, and imitate earlier adopters later in the cycle of innovation adoption. Probably the current rapid rate of technical evolution is partly because high rates of literacy, related phenomena such as the existence of libraries, and prosperity equip many people to make fairly effective individual decisions and to tolerate the cost of mistakes.

However, the technology of preindustrial societies can also be transformed quite rapidly. Among the most dramatic examples are those in which population growth leads to increased competition for resources. Exponential population growth is a very rapid process, and when it drives cultural change those processes can be rapid as well. For example, according to Kirch (1984) the Polynesian Islands were apparently settled by small groups of voyagers, and it took several hundred years for the population of larger islands to become overcrowded. However, as populations did reach these levels, considerable environmental deterioration occurred. At the same time, dense populations stimulated considerable technical innovation. On Hawaii, for example, irrigation, sophisticated dry-land farming systems, and aquiculture on a considerable scale allowed for an intensification of production in response to rising populations and deteriorating resources. Although status competition between chiefs (who supervised the larger-scale economic enterprises) played an important role in the evolution of late Polynesian technology, the basic decision-making forces of direct bias and guided variation must have been major elements of the process of invention and diffusion of the technology of intensification in response to more competition for a diminishing stock of traditional resources.

CULTURE IS AN ADAPTATION

If the massive use of culture is a distinctive human trait, what adaptive role does it play and how did a capacity for it evolve? In this section we try to understand the circumstances under which culture is superior to genetic transmission and individual learning as a means of adaptation. Understanding culture as an adaptation is important because we argue that some cultural processes may lead to maladaptive outcomes, but only if guided variation and direct bias are weak forces. Thus, it is important to know under what conditions selection might favor a strong reliance on social learning as opposed to personal experience. When does a tendency to depend on an inherited tradition become important relative to genetically inherited patterns of behavior, or a combination of genes plus individual learning?

Social learning is similar to both individual learning and genetic inheritance. Individual learning is a pure system of phenotypic adaptation to environmental contingencies, but the acquired adaptation perishes with the individual learner. A pure system of inheritance (genetic or cultural) does not allow the individual any flexibility. In this case, adaptation of the population results as a consequence of selection operating on heritable variation. Social learning allows both modes of adaptation. This mixed mode of adapting has two distinct advantages.

First, social learning may be favored because it allows individuals to avoid costs associated with learning. Individual learning may often be costly; it takes time, energy, exposes the organism to risk, and may require a larger brain. Alan Rogers (1989) has analyzed a simple recursion model in which there are two types of individuals—individual learners who evaluate alternative behaviors and choose the best one, and social learners who copy the behavior of a randomly chosen individual from the previous generation. He assumes that occasionally the environment changes from generation to generation. If individual learning is more costly than social learning, then through generations of the recursion model social learners always increase in frequency when they start out rare. This is because, surrounded by individual learners, they are virtually certain to acquire the best behavior without bearing the costs of individual learning themselves. However, as social learners become more common, fewer people are learning for themselves and errors begin to accumulate in the population. People who merely copy have a greater chance of copying another social learner, and thus acquiring an inferior behavior learned in a different environment. Rogers shows that at equilibrium of the recursion model there is always a mix of social and individual learners—the greater the environmental variability the lower the frequency of social learning. This recursion model also has the property that the average fitness of the equilibrium mix of social and individual learners is the same as a population composed only of individual learners. In this case, culture is favored by selection, but it does not increase average population fitness because the social learners are simply parasites on the efforts of those who do learn for themselves.

Second, social learning may be favored because it allows individuals to avoid learning errors. Virtually all learning mechanisms allow the possibility of error. Consider, for example, an individual trying to decide which of two foraging techniques is better. She tries them both out, and chooses the one that yields the highest return. Because yields will vary for many reasons, her trial may often yield erroneous results—the technique with the higher return during the trial may have a lower return over the long run. Costs and errors may be linked, because making learning more sophisticated and costly tends to reduce errors. Social learning can reduce the importance of such errors by allowing individuals to be more selective in their use of learned information. A social learning forager can use a rule such as this: Try out the two techniques and if one yields twice as much as the other adopt that technique, otherwise use the technique that mom used. The use of such a rule reduces the number of learning errors. However, it also slows the rate at which the population adapts. We have analyzed a simple recursion model that incorporates this idea. It suggests that, at equilibrium, individuals always depend on a mix of social and individual learning, and the average dependence on social learning increases as the environment becomes less variable in either time or space (Boyd and Richerson, 1988b, 1989a). Unlike Rogers's model, the equilibrium population may have higher average fitness than a population that depends only on individual learning; the cultural system for the inheritance of acquired variation is adaptive in changing environments.

So far, we have ignored genetic adaptation. We have seen that cultural inheritance is favored as environments become less variable. However, these are exactly the conditions under which selection will allow a population to adapt genetically to changing environments. What if we compare a system of genetic transmission plus individual learning, with a system of cultural transmission plus individual learning, for the same subsistence trait? We have done such an analysis using a recursion model conceptually similar to the one just described (Boyd and Richerson, 1985). It shows that the cultural system that can pass on acquired variation is favored over the system of genetic inheritance plus individual learning unless the environment is either nearly constant or nearly random. In the context of this model, the range of environments under which culture should be favored is rather broad.

These results give as much support as a simplified theoretical model can to our intuitive argument. A cultural system of inheritance combining individual and social learning ought to provide adaptive advantages to a population living in environments with an intermediate degree of environmental similarity from generation to generation. This is the regime where the evolutionary force of cumulative, relatively weak, low-cost individual learning pays off most. Individuals can depend primarily on tradition, yet the modest pressure of individual learning is sufficient to ensure that culture remains adapted to the moderately changing environment.

Existing data do not allow a critical empirical test of this adaptive rationale for culture. Two lines of inquiry might be pursued to develop such tests. First, many animals seem to have simple capacities for social learning; the best studied cases are Norway rats (Galef, 1988) and feral pigeons (Lefebvre and Palameta, 1988). These are weedy generalists that certainly have to adapt to variable environments. If a broader comparative study of animal social learning showed a significant correlation between environmental variability and capacities for social learning, the models would be supported. Second, humans are an extreme example of encephalization (brain enlargement), but many other animal lineages show more moderate encephalization during the Tertiary Period (the last 65 million years) (Jerison, 1973; Eisenberg, 1981). At least the last 2 million years of the Pleistocene seem to have much more variable climates than the past (Shackleton and Opdyke, 1976). It is perhaps not a coincidence that highly cultural hominids arose during the Pleistocene. The beginnings of the enlargement of the **neocortex** to the contemporary human scale began about the beginning of the Pleistocene as *Australopithecus* gave way to *Homo*, and fully *sapiens*-sized brains only evolved during the latter part of the Pleistocene with its high-amplitude glacial cycles (Klein, 1989). Studies of patterns of paleoclimatic variation are not yet sufficiently detailed to know if the relationship between increasing encephalization and increasing environmental variability is a close one on geological time scales across the broad spectrum of encephalizing lineages or not. If large brains are used for learning and social learning, the models suggest that increasing rates of environmental variation should have driven the Tertiary encephalization trends.

CULTURE IS MALADAPTIVE

Thus far we have seen how the forces of guided variation and direct bias can cause cultural evolution to mimic the results of genetic evolution. In such cases, it will be possible to predict behavior based on fitness maximization, even though the proximal cause of behavior is entirely culturally transmitted.

Other forms of cultural adaptation are not so simple. In this section we argue that there are processes of cultural adaptation that (1) lead to different outcomes than would be predicted based on fitness maximization, but (2) these are nonetheless favored by selection because they make social learning more effective.

Natural selection can act on cultural variation to produce evolutionary change in the same way that it acts on genetic variation. For natural selection to occur there must be variation, and the variants must differ in ways that affect the number of copies of each variant present in the next generation. Many culturally transmitted traits have substantial effects on fitness. Belonging to a pronatalist religion tends to increase fecundity, and belonging to an abstemious one tends to increase survival. Thus, if religious beliefs are transmitted from parents to offspring, selection on cultural variation can produce adaptations in the metric of genetic fitness.

People often acquire beliefs and values from individuals other than their parents. Such *asymmetric* cultural transmission makes adaptive sense. Direct bias is more effective if naive individuals survey many experienced others before they make up their minds whose trait to adopt. If dad is a poor hunter, why not observe the talents of several other men before making up your mind how you will approach the problem? Thus, selection may favor a tendency to imitate nonparental individuals.

When cultural variation is transmitted nonparentally, selection may favor genetically maladaptive cultural variants. Whenever individuals are culturally influenced by grandparents, teachers, peers, and so on, natural selection acting on cultural variation can favor those behaviors that increase the chance of attaining such nonparental roles. When the traits that maximize success in becoming a parent differ from those that maximize success as a teacher, priest, or grandparent, natural selection acting on cultural variation can cause genetically maladaptive traits to spread.

Demography and Reproduction

Human demography may provide important contemporary examples of this effect. Many urban populations, especially elite populations, throughout recorded history seemed to have reproduced at rates below what would be necessary for population replacement despite an economic capacity to out-reproduce nonelites (Knauft, 1987). Ancient cities were like tar pits, drawing country folk into their alluring but disease-ridden precincts. Princeton demographer Ansley Coale and his European Fertility Project colleagues have collected some examples of strong **fertility** control and population declines among rural populations in Europe before the 19th century, but these are rather isolated cases, because only urban societies could draw enough cultural recruits to sustain low-fertility behavior given premodern rates of mortality. Other fertility-controlling subcultures seem to have simply gone extinct.

The modern demographic transition that occurred in Europe at various times from the beginning of the 19th century onward altered and amplified the classic demographic effect of cities. As people have become wealthier in the industrialized countries, they have tended to lower their fertility, reversing on a society-wide scale the correlation between wealth and reproductive success often found in rural pastoral and agricultural societies.

It is plausible that low-fertility norms spread during the demographic transition in Europe because these norms led to cultural success. The modernization of Europe greatly increased the social complexity of European societies. Many new professions arose that were allocated on the basis of achievement, rather than on inherited rank as was common in premodern Europe. Professional educators and professional entrepreneurs are two examples. People in many of these new professions had an important role in education. Teachers served the need for universal literacy, and entrepreneurs and business managers trained factory workers, clerks, and other employees for the new occupations of the industrial era. Thus, relative to the agrarian society, European modernization must have resulted in more nonparental transmission from individuals in professional or entrepreneurial roles.

Modern data suggest that raising children who are likely to be successful in competition for professional or entrepreneurial roles conflicts with having a large family (Terhune, 1974). Children who do well in school and acquire similar skills for professional and entrepreneurial competition require considerable parental investment. It seems likely that the occupants of the new competitive roles tended to be drawn from a fraction of the population that already valued greater investments of parental effort per child, and therefore, all other things being equal, smaller families. As the weight of teachers in cultural transmission increased, and as these roles became allocated on the basis of achievement rather than inheritance, low-fertility norms could spread to the whole population. Thus, the spread of low-fertility norms may be one example of how natural selection acting on culturally transmitted variation may lead to behavior patterns that reduce individual reproductive success.

Why Selection Allows Culture to Deviate

The fact that natural selection may favor genetically maladaptive traits does not mean that cultural evolution necessarily results in maladaptive outcomes. Many core beliefs and values are usually acquired from parents and other close relatives, and selection on variants so transmitted tend to favor genetic fitness-maximizing behavior. Even if selection acts to increase the frequency of maladaptive beliefs, the effect may be unimportant if direct bias and guided variation are sufficiently powerful to keep genetically maladaptive variants at low frequency. Suppose, for example, natural selection acting on cultural variation favors a belief in a god who rewards the pious in heaven, and that this belief causes people to have smaller families than the genetic optimum, either because they join religious orders, or because they devote resources to support the church. Family sizes still might be optimal if the effect of selection is counteracted by direct bias. The evolved predispositions that underlie direct bias (for example, sexual desire and a love for children) might cause people to reject these religious beliefs, or at least to obey them mainly in the breach.

On the other hand, direct bias and guided variation may often fail to counteract the effects of selection on nonparentally transmitted variation because it is too difficult to determine which beliefs best serve the individual's genetic interest. Earlier we showed that when it is difficult to determine which of two variants is best, learning is costly and error prone, and, therefore, natural selection acting on genes favors a heavy reliance on cultural transmission. Thus, people may not reject religious beliefs because it is difficult to determine whether God exists. If He does, and if He rewards the pious, some evolved predispositions — fear of death, love of comfort — may overbalance the desire for a large family and lead people preferentially to adopt the practices sanctioned by religion. But determining whether God exists and exactly what He (She, They?) expect(s) of us has proven to be very difficult over the millennia.

The idea of a god who rewards the pious is only an especially striking example of a much larger class of cultural variation about which it is difficult and costly to make adaptive choices on the basis of evolved predispositions. The natural world is complex, hard to understand, and variable from place to place and time to time. Is witchcraft effective? What causes malaria? What are the best crops to grow in a particular location? Are natural events affected by human pleas to their governing spirits? The relationship between cause and effect in the social world is often equally hard to discern. What sort of person(s) should one marry? What mixture of devotion to work and family results in the most happiness or the highest fitness? People can make some intelligent guesses about such decisions, but the number of alternatives we can investigate in any detail is quite limited. Even if individuals are willing to devote substantial effort to particular decisions, each of us faces too many decisions to make costly investigations concerning all of them. The picture that emerges from behavioral decision theory is that people commonly rely on simple, often misleading, rules of thumb to make complex decisions (Nisbett and Ross, 1980). Human decision-making skills seem empirically to be a compromise between the rewards of accurate judgments, and the costs imposed by enlarging the cognitive apparatus and increasing the information collected from the environment.

As the effect of direct bias and guided variation weakens, culture becomes more

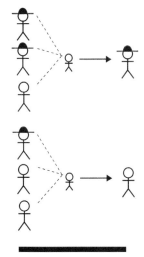

FIGURE 6.4

A schematic representation of frequency-dependent bias. Again there are two cultural variants: wearing a hat and going bare headed. Here we assume that young people acquire their sartorial beliefs by observing the behavior of three models, and that they are predisposed to imitate the more common behavior among the models they observe. This transmission rule increases the frequency of the more common type, in this case the habit of wearing a hat.

and more like a system of inheritance. Much of an individual's behavior is a product of beliefs, skills, ethical norms, and social attitudes that are acquired from a set of other people by social learning. To predict how individuals will behave, one must have knowledge about their cultural milieu. This does not mean that the evolved predispositions that underlie individual learning become unimportant. Without them cultural evolution would be uncoupled from genetic evolution, and would provide none of the fitness-enhancing advantages that must have favored the evolution of capacities for culture. However, it is also likely that cultural variation often responds to selection for behaviors that conflict with genetic fitness. Selection on genes that regulate the cultural system may still favor cultural transmission because on average it does better than genes could do alone.

These ideas are consistent with the view that much behavioral variation both within and among societies is genetically adaptive. We propose that people strive to satisfy evolved goals, but in the context of culturally acquired beliefs. Thus, *if the cultural context is taken as given, we would expect that much variation in behavior would be explicable in sociobiological terms.* For example, we would expect that a believer's decision to join a monastery is influenced by what he or she must give up in order to do so. A wealthy woman may well be more likely to enter a convent than a poorer one if it is customary to marry hypergynously, even if the religious beliefs of the rich and the poor are equally fervent. The same argument applies to variation among societies. A man may be more likely to become a celibate priest in a society where he can enhance his relatives' reproductive success because celibates are admired, wealthy, or powerful than in one in which such people are recruited by poor, despised religious minorities that restrict opportunities for nepotism.

More generally, we think it is plausible to view genetic and cultural evolution as a tightly coupled coevolutionary process. In some cases, forces such as guided variation and direct bias are strong and favor the spread of fitness-enhancing ideas and retard the spread of deleterious ideas. In such cases, fitness is increased by the presence of the cultural system, and behavior may be accurately predicted without explicit reference to the dynamics of cultural evolutionary processes. In other cases, the decision-making forces are weak, or expressed in a cultural environment that distorts their effects. In these cases, even when the cultural system does act to increase fitness, it is necessary to account for cultural effects in more detail.

Frequency-Dependent Bias

A similar argument can be made for a second bias force, **frequency-dependent bias**. This form of biased transmission allows individuals to improve their chance of acquiring adaptive behavior, but at the same time gives rise to processes that may not always result in fitness-maximizing behavior.

Guided variation, direct bias, and natural selection on parentally transmitted variation tend to cause adaptive behaviors to be more common than alternative behaviors. Thus, when it is difficult or costly to determine which variants are adaptive, it may be best to bias imitation in favor of the commonest type in the population. Recall the aphorism "When in Rome, do as the Romans do." We label this process *frequency-dependent bias* (Figure 6.4). Humans are widely suspected of conformity in their behavior, and this bias rule is quite plausibly important.

Frequency-dependent bias may cause **group selection** to be a more important process in cultural evolution than it seems to be in genetic evolution. Consider a large population subdivided into many smaller, partially isolated groups. Frequency-dependent bias reduces variation within groups because rarer variants are less likely to be imitated and therefore to become even rarer. At the same time, frequency-dependent bias increases variation between groups because immigrants entering a group are rare and also subject to discrimination in cultural transmission. Our theoretical work suggests that group selection may be more important in shaping cultural variation than it is in shaping genetic variation for this reason. If so, group-level adaptations may be more common in the human species than in other species.

Human societies exhibit much more cooperation than is typical of vertebrate societies. This tendency is most marked in the complex societies of the last few millennia, where the degree of division of labor, amount of altruistic self-sacrifice, and coordination of complex activities rival and exceed that of the advanced social insects. Even the simplest hunting and gathering societies are much more complex and cooperative than the societies of any other social mammal. The human sexes cooperate in an extensive division of labor between hunting and gathering. There is much sharing of food and other resources, especially within bands. Relatively peaceful, cooperative relations are generally maintained between several bands that share a common language and culture, numbering a few hundred to a few thousand individuals. By contrast, even among our closest relatives, the chimpanzees, the sexual division of labor is absent, food sharing (other than mothers with offspring) is minimal, and political cooperation is restricted to the handful of closely related males that form the core of a troop (Goodall, 1986).

In other animal societies, patterns of cooperation are well explained by **kin selection** and **reciprocal altruism**. For example, the complex societies of the social insects are based on kin selection. Only a few closely related individuals are reproductively active in the colony, and the cooperating sterile workers are their offspring (Wilson, 1975). The best documented cases of reciprocal altruism involve pairs of individuals. For example, vampire bats exchange regurgitated blood reciprocally in pairs (Wilkinson, 1984).

It is an open question whether either of these two mechanisms is sufficient to explain the scale of cooperation observed in even the smallest-scale human societies. Explanations based on kin selection cannot easily explain why humans cooperate with nonrelatives in large-scale societies. However, Pierre van den Berghe (1981) has proposed that kin selection accounts for patterns of cooperation observed in small-scale societies, and that cooperation in complex societies is the result of a cognitive mistake. The empirical problem with reciprocal altruism is that there are no known cases of large-scale cooperation attributable to reciprocity (unless humans are such). Thus, Richard Alexander (1987) has argued that our uniquely extensive cooperation is supported by complex webs of "indirect reciprocity" that are restricted to humans because only our species has the cognitive sophistication to keep track of the large number of interactions that result.

Group selection based on cultural variation is also a possible explanation for the evolution of human cooperation. Frequency-dependent bias may maintain enough cultural variation among groups for group selection to be important. It has the by-product of discriminating against rare variants in the population. Thus, a fair amount of immigration of less altruistic individuals does not convert a more altruistic group to a less altruistic one. As long as the less altruistic variants are a minority, the conformity effect acts as a powerful impediment to this variant's increase despite assuming the usual within-group advantage to less altruistic behavior. This mechanism does not even require the demographic annihilation of groups with too few altruists, merely their disruption and the dispersal of members. So long as it is rare for such dispersal events to tip more altruistic populations over the threshold where the less altruistic variant begins to increase, selection between groups can be a potent force (Boyd and Richerson, 1990).

None of these hypotheses about altruism is completely implausible. Our theoretical studies of reciprocity in large groups, including models incorporating Alexander's idea of indirect reciprocity and models of punishment, suggest that reciprocity should be restricted to quite small groups (Boyd and Richerson, 1988a, 1989b, in press). Even in groups as small as six to ten, reciprocity is much more difficult to start in a population than it is when only pairs interact. The models also indicate that a synergistic combination of kin selection and reciprocity does not tend to make reciprocity easy to start when it is rare in larger groups, unlike the case for pairs. However, this area is still poorly explored, and perhaps strategies as yet unexamined by recursion models will be more effective. It is certainly possible that some combination of kinship and reciprocity can explain cooperation in small-scale, face-to-face societies, but not in larger scale agricultural and industrial ones. Donald Campbell (1983) argues that cultural group selection is only necessary to explain the levels of cooperation and integration in the complex, large-scale societies of the last few thousand years. These societies are so large as to involve extensive cooperation among largely anonymous masses of people. It is harder to see how kin altruism and reciprocity can knit these societies into more workable complexes than hunting and gathering societies where much of the political power may reside in a hundred or so adult males, who cooperate for the most part in coresidential groups of only ten or so. Critical theoretical and empirical work is only beginning on this important problem.

■
SUMMARY

In this chapter we argue that cultural evolution is a Darwinian process. Culture is like genes in the sense that information about how to behave is transmitted from individual to individual. Each individual samples the culture of the past by observing others or by being taught, and then potentially becomes sampled in turn. But in many other respects, culture is unlike genes. One or many cultural "parents" may be sampled instead of only two, for example. The most fundamental structural difference between genes and culture is that cultural inheritance is a system for the inheritance of acquired variation. Individuals' capacity for learning and decision making is harnessed directly to the cultural transmission system in ways that apparently do not exist in the case of genes.

The differences between the genetic and cultural systems give rise to interesting scientific problems. How does the cultural evolutionary process work? How does it interact with the genetic evolutionary process to produce adaptations? What are we to make of apparently maladaptive cultural practices?

The most important difference between the evolutionary processes of the genetic and cultural systems is the existence of decision-making forces in the cultural system. Not only does the survival and reproduction of variant individuals cause evolutionary change in the cultural system. The decisions that people make as they learn for themselves or decide whom to imitate or what behaviors to adopt also affect cultural evolution.

The decision-making forces—guided variation and the various forms of biased transmission—give the cultural system of inheritance an adaptive advantage in certain kinds of variable environments. When individual learning is coupled with the possibility of transmission by imitation, the cultural system can track environmental fluctuations more quickly than can genes, and at a lesser information cost (or with fewer errors) than relying entirely on individual learning. It is optimal to depend mostly on imitation when learning is costly or error prone, and when the environment does not change too rapidly. Even in this case, the small amount of individual learning is important; it causes the population to track the changing environment more effectively than genes, and can give the cultural inheritance system a considerable advantage over the more familiar genes-plus-individual-learning system. Human diet choices, for example, may well be closer to optimal than they otherwise would be due to this effect.

In addition to making ordinary adaptive processes more efficient, the existence of culture may have contributed to qualitatively new human adaptations. Human **eusociality** is an example. We hypothesize that cultural processes like conformist transmission permit a measure of group selection on cultural variation. The food sharing and division of labor of hunting and gathering bands appears to have been crucial to the extraordinary geographical expansion of the human species during the Pleistocene. The demographic success of complex societies is clearly dependent on cooperation and the division of labor. Culture may be the analog of the peculiar **haplodiploid** system of sex determination in the ants, bees, and wasps that makes sisters more related to each other, and hence prone to the evolution of sterile-worker eusociality. Conformist transmission can raise cultural relatedness far above genetic relatedness, even in large groups. Note that in a group in which the cultural environment has evolved to favor altruistic behavior, genetic impulses to altruism might be favored by mate selection or other socially conferred sanctions and rewards. Those with a genetic predisposition to altruism may have had greater mating success, and a higher standing in the community attended by greater survival.

Finally, there is no guarantee that all cultural traits will be adaptive from the genetic point of view. The existence of nonparental transmission, among other things, gives culture a measure of evolutionary activity in its own right. When the decision making that might more closely control cultural evolution is costly, genetic fitness is best served by a system that tolerates some deviance from genetic fitness optimization. Better some systematic cultural deviations from fitness optima than more severe random ones due to individual error.

A quite suggestive case can be made that the theoretically arresting cases of novel human adaptations and unique kinds of maladaptations due to culture are also real and important. But only suggestive! The amount of critical field work and experimentation that has been undertaken to test these ideas is still quite small. We as yet know far less about cultural evolutionary processes and their interactions with the genetic system than we know about ordinary organic evolution. It is an interesting historical paradox that we know least about evolutionary processes in the animal whose evolution interests us most.

■

SUGGESTED READINGS

Boyd, R., and Richerson, P. J. 1985. *Culture and the Evolutionary Process* (Chicago: Univ. of Chicago Press).

Boyd, R., and Richerson, P. J. 1987. The evolution of ethnic markers. *Cultural Anthropology* 2: 65–79.

Campbell, D. T. 1975. On the conflicts between biological and social evolution and between psychology and moral tradition. *American Psychologist 30*: 1103–1126.

Cavalli-Sforza, L. L., and Feldman, M. W. 1981. *Cultural Transmission and Evolution: A Quantitative Approach* (Princeton, NJ: Princeton Univ. Press).

■

REFERENCES

Alexander, R. D. 1987. *The Biology of Moral Systems* (New York: Aldine de Gruyter).

Boyd, R., and Richerson, P. J. 1985. *Culture and the Evolutionary Process* (Chicago: Univ. of Chicago Press).

Boyd, R., and Richerson, P. J. 1988a. The evolution of reciprocity in sizable groups. *Journal of Theoretical Biology 132*: 337–356.

Boyd, R., and Richerson, P. J. 1988b. The evolution of social learning: The effects of spatial and temporal variation. In: T. R. Zentall and B. G. Galef, Jr. (Eds.), *Social Learning: Psychological and Biological Perspectives* (Hillsdale, NJ: Lawrence Erlbaum), pp. 29–48.

Boyd, R., and Richerson, P. J. 1989a. Social learning as an adaptation. *Lectures on Mathematics in the Life Sciences 20*: 1–26.

Boyd, R., and Richerson, P. J. 1989b. The evolution of indirect reciprocity. *Social Networks 11*: 213–236.

Boyd, R., and Richerson, P. J. 1990. Group selection among alternative evolutionarily stable strategies. *Journal of Theoretical Biology 145*: 331–342.

Boyd, R., and Richerson, P. J. In press. Punishment allows the evolution of reciprocity (or anything else) in sizable groups. *Ethology and Sociobiology*.

Campbell, D. T. 1983. The two distinct routes beyond kin selection to ultrasociality: Implications for the social sciences and humanities. In: D. Bridgeman (Ed.), *The Nature of Prosocial Development: Theories and Strategies* (New York: Academic Press), pp. 11–39.

Cosmides, L., and Tooby, J. 1989. Evolutionary psychology and the generation of culture, II. Case study: A computational theory of social exchange. *Ethology and Sociobiology 10*: 51–97.

Eisenberg, J. F. 1981. *The Mammalian Radiations: An Analysis of Trends in Evolution, Adaptation, and Behavior* (Chicago: Univ. of Chicago Press).

Galef, B. G., Jr. 1988. Communication of information concerning distant diets in social, central-place foraging species: *Rattus norvegicus*. In: T. R. Zentall and B. G. Galef, Jr. (Eds.), *Social Learning: Psychological and Biological Perspectives* (Hillsdale, NJ: Lawrence Erlbaum), pp. 119–139.

Goodall, J. 1986. *The Chimpanzees of Gombe: Patterns of Behavior* (Cambridge, MA: Harvard Univ. Press).

Jerison, H. J. 1973. *Evolution of the Brain and Intelligence* (New York: Academic).

Kirch, P. 1984. *The Evolution of Polynesian Chiefdoms* (Cambridge: Cambridge Univ. Press).

Klein, R. G. 1989. *The Human Career: Human Biological and Cultural Origins* (Chicago: Univ. of Chicago Press).

Knauft, B. M. 1987. Divergence between cultural success and reproductive fitness in preindustrial cities. *Cultural Anthropology 2*: 94–114.

Lefebvre, L., and Palameta, B. 1988. Mechanisms, ecology, and population diffusion of socially learned, food-finding behavior in feral pigeons. In: T. R. Zentall and B. G. Galef, Jr. (Eds.), *Social Learning: Psychological and Biological Perspectives* (Hillsdale, NJ: Lawrence Erlbaum), pp. 141–164.

Lumsden, C. J., and Wilson, E. O. 1981. *Genes, Mind, and Culture: The Coevolutionary Process* (Cambridge, MA: Harvard Univ. Press).

Nisbett, R., and Ross, L. 1980. *Human Inference: Strategies and Shortcomings of Social Judgment* (Englewood Cliffs, NJ: Prentice-Hall).

Rogers, A. R. 1989. Does biology constrain culture? *American Anthropologist 90*: 819–831.

Rogers, E. M. 1983. *Diffusion of Innovations* (New York: Free Press).

Shackleton, N. J., and Opdyke, N. D. 1976. Oxygen isotope and paleomagnetic stratigraphy of Pacific core V28–239, Late Pliocene to Latest Pleistocene. *Geological Society of America Memoir 145*: 449-464.

Smith, E. A., and Winterhalder, B. 1992. *Evolutionary Ecology and Human Behavior* (New York: Aldine de Gruyter).

Terhune, K. W. 1974. A review of the actual and expected consequences of family size. *Calspan Report N. DP-5333-G-1, Center for Population Research. National Institute of Child Health and Human Development, U.S. Dept. of Health Education and Welfare Publication* (*NIH*), pp. 75-779.

Tooby, J., and Cosmides, L. 1989. Evolutionary psychology and the generation of culture. *Ethology and Sociobiology 10*: 29-50.

Van den Berghe, P. L. 1981. *The Ethnic Phenomenon* (New York: Elsevier).

Wilkinson, G. 1984. Reciprocal food sharing in vampire bats. *Nature 308*: 181-184.

Wilson, E. O. 1975. *Sociobiology: The New Synthesis* (Cambridge, MA: Harvard Univ. Press).

Zentall, T. R., and Galef, B. G., Jr. 1988. *Social Learning: Psychological and Biological Perspectives* (Hillsdale, NJ: Lawrence Erlbaum).

GLOSSARY

Abductor. Muscle or muscle group that *abducts,* moves a limb away from the body axis, or a digit away from the limb axis.

Accelerator dating. A form of *radiocarbon dating* that measures extremely minute quantities of *carbon 14* by use of an accelerator, a device that imparts high velocities to charged particles thereby separating *isotopes* by mass.

Acheulian. Stone tool "industry" of the Middle and part of the Lower Pleistocene, usually associated with *Homo erectus.*

Allele. Any of two or more alternative forms of a gene at the same *locus* (position) on a *chromosome.*

Altricial. Having the young born in a very immature and helpless condition requiring care for some time.

Amenorrhea. Absence or temporary suspension of the menstrual cycle reflecting an inactive ovarian cycle, one effect being the cessation of *ovulation.*

Anatomically modern *Homo sapiens*. Informal taxonomic grouping of those populations having high foreheads, distinct chins, reduced or absent brow ridges, and other anatomical features characteristic of all living human populations.

Archaic *Homo sapiens*. Informal taxonomic grouping of hominids anatomically intermediate in some respects between *Homo erectus* and *Homo sapiens,* often characterized by brain size larger than that of *Homo erectus* but retaining prominent *supraorbital* ridges or *tori,* reduced foreheads, and no distinct chin; the precise taxonomic status of these populations is uncertain.

Archeology (or *archaeology*). Branch of anthropology concerned with past human behavior, culture and ecology as revealed by artifacts, occupation areas, plant and animal parts, and other remains; compare *paleontology.*

Argon dating. Methods that measure *isotopes* of potassium and the stable gas argon to date volcanic rock: (1) potassium-argon dating uses the known rate of radioactive decay of the potassium isotope ^{40}K to the inert gas argon (^{40}Ar) and calcium-40 (^{40}Ca); (2) in single crystal ^{39}Ar-^{40}Ar dating, the rock sample is irradiated to convert ^{39}K to ^{39}Ar, and then heated to near its melting point to release argon gas from the interior of mineral grains, thereby allowing the ratio of ^{40}Ar to ^{39}Ar in the gas to be measured.

Articular surface. Portion of a bone that attaches to (articulates with) another bone through an intervening joint of cartilage.

Aurignacian. A material culture period of the early Upper Paleolithic in Europe, associated with Cro-Magnon people.

Australopithecine. Informal name for a member of the genus *Australopithecus;* also, of or related to such.

Australopithecus. Genus name derived from *australo-* (southern) and *-pithecus* (ape, or literally, "one who plays tricks") applied to hominid species characterized by relatively small body size, relatively high ratio of forelimb to hindlimb length, ape-sized brain, *megadontia,* and varying degrees of hypertrophy of the chewing musculature and associated cranial structures.

Australopithecus afarensis. The earliest known species of hominid; a small-bodied, gracile species from East Africa.

Australopithecus africanus. Gracile hominid species similar to *A. afarensis* found in limestone cave deposits of South Africa.

Australopithecus boisei. A hominid species from East Africa adapted to grinding resistant food items, as indicated by robust development of the *mandible,* lower face, ridges of muscle attachment on the *cranium,* and relatively huge molars and premolars.

Australopithecus robustus. A robust-skulled hominid species similar to *A. boisei* found in limestone cave deposits of South Africa.

Axilla. Polite name for armpit.

Behavioral ecology. Discipline that examines how behavior has evolved, and how it may be adaptive, in a particular ecological context.

Beta globin. A particular chain or segment of a colorless blood protein obtained by removal of a "heme" (a deep red iron-containing pigment) from a "hemoglobin" (a red blood cell molecule functioning in oxygen transport).

Biased transmission. Process of cultural evolution whereby individuals choose from among several alternatives which *cultural variants* they will adopt, or which members of society they will imitate.

Biface. Stone artifact with flakes removed from two intersecting surfaces, resulting in a sharp edge.

Binder. Substance added to paint or other fluids to improve qualities of texture and adhesion.

Biotite. A generally black or dark-green form of mica forming a constituent of crystalline rocks, and consisting of a silicate of iron, magnesium, potassium, and aluminum.

Bipedalism. Two-footed posture and locomotion, especially referring to the human attribute of upright stance on the hindlimbs.

Blood groups. Any one of several *polymorphisms* of blood serum proteins; for example, the ABO blood group.

Bonobo. A species of hominoid, *Pan paniscus,* that lives in forested habitats of central Africa south of the Zaire River; the closest living relative of the common chimpanzee, *Pan troglodytes.*

Broca's area. A *cortical* region of the human brain located on the side of the frontal lobe, just above the temporal lobe (directly beneath a finger placed at the temple). This area has long been implicated in the production of speech because injury to it will result in aphasias (language dysfunction).

C₃ woody plants. Those plants, including most tropical woody species, that use a photosynthetic pathway that absorbs a moderate level of the *stable isotope* of carbon, ¹³C; compare *C₄ grasses.*

C₄ grasses. Those plants, including most tropical grasses, that use a photosynthetic pathway that absorbs a high level of ¹³C; compare *C₃ woody plants.*

Callitrichid. A member of the primate family Callitrichidae, containing the marmosets and tamarins of South America; also, of or related to such.

Calvaria (plural *calvariae*). The brain case or cranial vault lacking the mandible, the face, and often the basal portions of the skull.

Canine. Structurally simple, conical tooth located between the *incisors* and *premolars;* one in each quadrant of the mouth (upper and lower, right and left); the dentist's eye tooth.

Carbon 14 (¹⁴C). A radioactive *isotope* of carbon produced in the atmosphere that occurs in living plants and animals at a particular ratio with ¹²C; it decays to nitrogen 14 (¹⁴N) and a beta particle with a half-life of 5,730 years.

Carbonate. A mineral containing the chemical group CO₃, such as calcite, or a rock consisting chiefly of such minerals, such as limestone.

Carpals. Small bones of the wrist and hand; in humans, the scaphoid, lunate, triquetrum, pisiform, hamate, capitate, trapezoid and trapezium; see inside back cover.

Cenozoic. Geological era spanning the time from about 65 million years to now; often characterized as the Age of Mammals for the preponderance of mammals preserved in the terrestrial fossil record, although many, many other groups (e.g., beetles and birds) have been even more diverse and abundant.

Cercopithecoid. A member of the superfamily Cercopithecoidea; also, of or related to such.

Cercopithecoidea. Zoological superfamily containing primates characterized by the combination of quadrupedal locomotion, only two premolars in each jaw quadrant, and bilophodont molars (two-lophed-teeth, teeth bearing two high crests or lophs). Although often called Old World monkeys, cercopithecoids are more closely related to hominoids than they are to New World monkeys; familiar examples are baboons, vervets, macaques, mangabeys, guenons, langurs and proboscis monkeys.

Cerebral. Of or related to the *cerebrum,* the front and upper portion of the brain, greatly enlarged in mammals.

Cervical. Of or pertaining to the neck, especially the seven neck *vertebrae;* see inside back cover.

Châtelperronian. A material culture period of the Upper Paleolithic in Europe, at or near the transition from *Neanderthal*-dominated to *Cro-Magnon*-dominated populations.

Chromosome. The gene-carrying structure within a cell nucleus that contains *DNA.*

Clade. A group of organisms descended from a single common ancestor; a monophyletic group.

Cladogram. A branching diagrammatic tree used to illustrate the sequence of evolutionary divergence within a group of organisms.

Clavicle. Collarbone; the bone connecting the *sternum* (breastbone) to the *scapula* (shoulder blade); see inside back cover.

Claviform. Shaped like a club.

Coccyx. Tailbone; the last few *vertebrae* of the *vertebral column,* often fused into a single unit; see inside back cover.

Consortship. In primatology, a close behavioral relationship between a male and an *estrous* female.

Conspecific. Member of the same species.

Continental plate. Any one of several autonomous structural units of the Earth's crust carrying continental land mass; over geologic time these plates move across the surface of the planet, a phenomenon called continental drift (a concept readily grasped by Californians).

Core. Stone that has been reduced by *flake* removal.

Cortex. The outer or peripheral portion of an organ or plant part; in animals, used especially in reference to the outer layer of the brain.

Cortical. Of or related to the *cortex.*

Cranium (plural *crania*). The skull minus the mandible; see inside back cover.

Cro-Magnon. Population of hominids that represents the earliest anatomically modern *Homo sapiens* known from the European Upper Paleolithic; named after the site of Cro-Magnon, near Les Eyzies, France.

Cultural variant. Any one of the many behavioral variations or options that occur within a human society, and which may be learned by other individuals.

Deciduous. Being shed seasonally or at a specific stage of development, especially referring to the first set of teeth that erupts during infancy, to be replaced by the adult set of teeth during childhood.

Demography. The statistical study of population size, density, mortality, and reproductive replacement.

Dimorphism. The condition of having two different forms or sizes, often used in reference to gender differences.

Direct bias. Form of *biased transmission* involving individual learning, in which individuals choose the *cultural variant* or the cultural model they prefer based on their own experience or motivations.

Distal. Far away from the central axis of the body.

DNA. Deoxyribonucleic acid; a large molecule that is the genetic material for most life on Earth. In structure, it comprises two parallel strands of sugars and phosphates, with cross-ties between strands formed of pairs drawn from four types of *nucleotide bases,* the whole structure helically wound.

Electron spin resonance. Dating method that detects the magnetic effects of negatively charged "spinning" electrons in an ionized substance; the ionization is created by low-level, natural environmental radiation and its magnetic effects build up over time, thus providing an estimate of age; compare *thermoluminescence.*

Enamel. Hard white substance that covers the entire surface of a tooth crown.

Enculturation. Process by which individuals acquire behavior, beliefs, and information typical of the culture in which they are raised or in which they live.

Endocast. An impression or cast of the inside of something, usually the brain case, providing a morphological replica of the originally enclosed soft tissue surface.

Endocranial volume. Volume of the cavity within a skull that in life houses the brain; slightly greater than brain volume, due to the presence of membranes, vessels, and fluid between brain surface and the internal skull surface.

Epigamic. Of or pertaining to secondary sexual features, visual or olfactory features distinctive of only one gender but not necessarily functionally related to the primary sex organs.

Epiphysis (plural *epiphyses*). A secondary center of *ossification* (bone formation) usually located at the ends of limb bones, separated from the primary center (usually the shaft) by a cartilage plate where growth occurs; the epiphyses fuse to the shaft when bone elongation is completed.

Estrus. A time of increased female sexual activity, often accompanied by enhancement of visual or olfactory sexual signals, that occurs at and around the time of *ovulation;* adjective form is *estrous.*

Ethnologist. Anthropologist dealing chiefly with the comparative and analytical study of living human cultures.

Eusociality. "True sociality," applied to species in which individuals live in groups, cooperate in caring for the young, and have a division of labor associated with reproduction; in some species, such as bees, ants, termites, and naked mole rats, the division of labor is developed to the point of having sterile worker castes that feed and raise offspring.

Extender. Substance added to a material such as paint in the capacity of a diluent or modifier.

Fecundity. Number of offspring a female is biologically capable of producing over a given interval or lifetime; compare *fertility.*

Femur (plural *femurs* or *femora*). Long bone of the thigh; see inside back cover.

Fertility. Number of offspring actually produced over a given interval or lifetime; compare *fecundity.*

Fibula. One of two long bones of the lower leg, this the more slender, lateral one; see inside back cover.

Fitness. Contribution of one individual's *genotype* to succeeding generations compared with the average of all others; or, the contribution of one *allele* to succeeding generations compared with the average of all others at that *locus.*

Fitness-optimizing theory. A theoretical branch of the biological and social sciences that tries to model or predict the adaptations of organisms by finding an optimal balance among various ecological costs and benefits yielding the greatest net production of successful offspring.

Flake. A usually sharp-edged stone fragment struck or pressured off of a larger rock.

Folivore. An organism that eats leaves and stems.

Frequency-dependent bias. Form of *biased transmission* in which an individual models his or her behavior after the most common *cultural variants* in the population.

Frugivore. An organism that eats fruit.

Gene flow. Geographic movement of genes caused by migration, dispersal, or mating patterns of organisms.

Gene pool. The total complement of genes in a population.

Genetics. Branch of biology that deals with heredity and variation of organisms.

Genome. The total genetic material of an organism.

Genotype. Set of genes possessed by an individual organism; compare *phenotype.*

Gigantopithecus. Hominoid genus containing two species of huge extinct apes: Miocene *G. giganteus* of India and Pakistan, and Pleistocene *G. blacki* of southeast Asia; estimated body size of the latter is as much as 300 kg.

Group selection. Evolutionary process involving differential survival and reproduction of groups, rather than of individuals.

Guided variation. Process of cultural evolution involving individual learning, by which individuals modify their own behavior, beliefs, or values as a consequence of individual learning or individual experience, thereby passing on new, modified versions of the behavior to others.

Hammerstone. Rock used to strike *flakes* from *cores,* usually identifiable by a local battered region on its surface.

Handaxe. A teardrop shaped, bifacial stone implement.

Haplodiploidy. System of sex determination found in some beetles, ants, bees, and other insects whereby fertilized (diploid) embryos develop into females and unfertilized (haploid) embryos become males, which has consequences for the degree of genetic relatedness among relatives; for example, sisters are more closely related to each other than they are to their own offspring.

Hematite. A naturally occurring mineral, the iron oxide Fe_2O_3, often used as a pigment.

Holocene. Current epoch of the Cenozoic Era; see inside front cover.

Hominid. A member of the family Hominidae; also, of or related to such.

Hominidae. Zoological family containing *bipedal* hominoids.

Hominoid. A member of the superfamily Hominoidea; also, of or related to such.

Hominoidea. Zoological superfamily containing gibbons and siamangs (the lesser apes), orangutans, chimpanzees, *bonobos,* and gorillas (the great apes), human beings, and also many extinct species.

Homo. Primate genus containing species of relatively small-toothed, large-brained, stone-tool-making hominids.

Homo erectus. An extinct species of relatively large African and Eurasian hominids characterized by a modern *postcranium,* slightly projecting face, pronounced *supraorbital* and *occipital torus,* and medium-sized brain (by hominid standards).

Homo habilis. A small species of hominid restricted to Africa at about the time of the Pliocene-Pleistocene transition; the earliest widely recognized member of the genus *Homo.*

Homo heidelbergensis. A formal species name that may apply to the material often referred to as archaic *Homo sapiens,* if this group really represents a distinct species; the *type specimen* is the Mauer mandible found near Heidelberg, Germany.

Homo neanderthalensis. See *Neanderthals.*

Homo sapiens. The species you belong to.

Homo sapiens sapiens. Subspecies or subpopulation of all modern humans plus those prehistoric populations that unambiguously express the full development of *anatomically modern* features.

Horizon. A particular stratigraphic level or time interval.

Humerus. Long bone of the upper arm; see inside back cover.

Hyoid. Small bone of the throat, site of muscle attachments, functioning in the control of fine movements of the lower jaw and throat; see inside back cover.

Hypervitaminosis A. Condition brought on by an excessive dietary intake of vitamin A, resulting in hemorrhage along bone surfaces, and subsequent irregular formation of bone there.

Hypodigm. All of the individual specimens that are believed to represent a particular species; compare *type specimen.*

Iliac blades. Broad, flat plates of the ilium, one of the bony components of the *pelvis;* the hip bones.

Incisor. Anterior teeth that in hominoids are broad and relatively flat-edged, often functioning to break food items into manageable pieces; there are two incisors per jaw quadrant (upper and lower, right and left).

Innominate bone. Large bone of the *pelvis;* see inside back cover.

Intercostal. "Between the ribs," especially used in reference to the internal and external intercostal muscles and their associated nerves and blood vessels; these muscles lift, expand, and compress the rib cage, thereby contributing to the control of breathing and other lung movements.

Isotope. Any of two or more types of atoms of a chemical element having nearly identical chemical behavior (same number of protons) but with differing atomic mass and physical properties (variable number of neutrons).

Karyotype. The number and structure of the *chromosomes.*

Kin selection. Evolutionary process involving differential reproduction of *alleles* as measured not just by the reproductive output of a given individual, but also that of all relatives that might carry a copy of the same allele; a mechanism by which cooperative behavior among related individuals can evolve.

Knuckle-walker. Quadrupedal animal that supports the upper body weight while traveling with the knuckles of the *phalanges,* such as gorillas and chimpanzees.

Lactation. The secretion of milk by the mammary glands.

Laetoli footprints. Foot impressions of many animals, including *bipedal* hominids, that were made on a broad surface of soft volcanic ash shortly after an eruption about 3.7 million years ago, buried and preserved by subsequent layers of ash, at the site of Laetoli, Tanzania; the earliest known record of bipedal hominids.

Lithic. Of or relating to stone.

Locus. Site on a *chromosome* occupied by a specific gene; more informally, the gene itself including all its *alleles.*

Lower Paleolithic. The earliest (stratigraphically lowest) division of the Old Stone Age, including the time encompassing stone tool cultures of hominid species other than *Homo sapiens;* see inside front cover.

Lumbar. Of or pertaining to the lower back, especially the vertebrae that lie between the rib cage and the pelvic girdle; see inside back cover.

Magdalenian. Final material culture period of the Upper Paleolithic in Europe, often broken down into early, middle and late subdivisions.

Mandible. Lower jaw; see inside back cover.

Manganese dioxide. Dark insoluble compound, MnO_2, used especially as an oxidizing agent.

Megadontia. The condition of having relatively large teeth.

Meiosis. Process by which the genetic material of a diploid cell (containing two copies of each *chromosome*) is reshuffled to make unique gametes (egg and sperm). Entails replication of the *DNA* and then division of the diploid cell nucleus into four haploid daughter cells (each containing one copy of each chromosome); during the process, portions of homologous chromosome pairs are exchanged (*crossing-over*), and nonhomologous chromosomes segregate independently, thereby generating great diversity of genetic combinations.

Mesozoic. Geological era spanning the time from about 250 to 65 million years; often characterized as the Age of Dinosaurs, but also including a great diversity of other animal and plant life, including early mammals; see inside front cover.

Metacarpal. One of five bones of each hand, lying between the *carpals* and the fingers.

Metatarsal. One of five bones of each foot, lying between the *tarsals* and the toes.

Midden. A refuse heap.

Middle Paleolithic. Division of the Old Stone Age, characterized by stone *flake* industries such as the *Mousterian* of Europe, associated with Neanderthals; see inside front cover.

Milk molar. A deciduous molar that is eventually replaced by an adult premolar; called "molar" because the crown structure resembles that of adult molars. No deciduous teeth precede the definitive adult molars.

Miocene. Geological epoch of the Cenozoic Era, 24 to 5 million years ago; characterized by climatic changes from warm, humid conditions in the early part of the epoch to cooler and drier conditions at the end; see inside front cover.

Mitochondrial DNA. *DNA* that occurs in a mitochondrium, as opposed to a cell nucleus; mitochondria are organelles that produce energy in animal cells. Two features of mitochondrial DNA (mtDNA) are of special interest: (1) it is transmitted through generations only in the female's egg, and (2) it does not sexually reproduce or undergo *meiosis.* Therefore, an individual's mtDNA represents a separate evolutionary lineage traceable through the maternal line and differing from the mtDNA of another individual only by the *mutations* that have accrued since their last common female ancestor.

Molar. Any one of the posterior teeth that in hominoids are large, broad teeth with multiple low cusps, functioning primarily for mastication of food; there are usually three molars in each quadrant of the mouth (upper and lower, right and left).

Monomorphism. The condition of having only one size or shape, usually in reference to sexual monomorphism, in which the two sexes are similar to each other.

Mousterian. European stone tool "industry" of the Middle *Paleolithic,* usually associated with Neanderthals.

Mutation. An error in replication or other alteration of the *nucleotide base* sequence of *DNA;* new heritable genetic variation.

Natural selection. The process of differential survival and reproduction (not due to chance) of classes of entities that differ in heritable characteristics; the entities may be *alleles,* individual organisms, kin groups, or populations. Usually applied to the case of individual organisms; compare *group selection, kin selection.* (Differential representation of heritable variation in subsequent generations due to chance is genetic drift.)

Neanderthals. Late Pleistocene population of the genus

Homo occurring in Europe and the Middle East; may represent a population of archaic *Homo sapiens,* designated *Homo sapiens neanderthalensis,* or may be a distinct species, *Homo neanderthalensis.*

Neocortex. The *cortex* of the cerebral hemispheres, seat of many sensory, motor, and higher mental functions.

Neural canal. The large opening through the *vertebrae* that contains the spinal cord; also called the vertebral canal.

Neural spine. A blade of bone projecting dorsally from a *vertebra,* serving as attachment site for several ligaments and muscles; also called a spinous process.

Neuron. Nerve cell.

Nucleotide. A subunit of *DNA* that consists of a purine or a pyrimidine nitrogen-containing *nucleotide base* combined with a sugar and a phosphate group; see *DNA.*

Nucleotide base. One of four purine or pyrimidine nitrogen-containing compounds (adenine, cytosine, guanine, thymine) that serve to bind the double strands of DNA together; the sequence of bases in a strand of nucleotides serves as the "alphabet" of the genetic code.

Obstetrical. Of or relating to childbirth.

Occipital torus (plural *tori*). A buttress or protuberance of bone (a torus) at the back of the head (the occipital region, the occipital bone); especially characteristic of *Homo erectus.*

Oldowan. Stone tool complex of the late Pliocene and early Pleistocene, usually associated with *Homo habilis* and early *Homo erectus,* although australopithecines may also have manufactured Oldowan tools.

Old World monkeys. Members of the superfamily Cercopithecoidea.

Omnivore. An organism that eats a diversity of food types.

Operant conditioning. Form of training or learning whereby an organism's actions or responses to the environment produce rewarding and reinforcing effects.

Opercular cortex. Folded surface layer at the lateral part of the brain, lying within and forming parts of the rim of the lateral sulcus, a large infolding of the *cortex* at the boundary between the frontal lobe and temporal lobe.

Ossification. The process of bone formation.

Osteodontokeratic culture. From the Greek roots *osteo-,* referring to bone, *-donto-,* to tooth, and *-keratic,* to horn, the name given to a hypothetical early human culture making tools of animal parts but not of stone.

Outgroup. A taxon or group of organisms that diverged from another group of taxa before the latter diverged from each other.

Ovulation. Release of an unfertilized gamete (egg) from the ovary (female reproductive organ that generates and stores eggs).

Paleoanthropology. The broad study of the human past, including contributions from archeology, paleontology, primatology, geology, and evolutionary biology.

Paleolithic. Old Stone Age; the material culture period beginning with the appearance of the *Oldowan* in the late Pliocene and extending to the end of the Pleistocene glaciation.

Paleontology. The science concerned with past life as studied from fossilized remains of organisms and geological evidence; compare *archeology.*

Palynology. The science concerned with the study of plant pollen and spores, especially in archeological and paleontological contexts.

Paranthropus. A generic name derived from *para-* (alongside, or closely related to) and *-anthropus* (human beings), first used for a South African species of robust australopithecine; considered by some to be a valid genus containing all robust australopithecine species, but considered by others to be unworthy of generic separation from *Australopithecus.*

Patella. Knee cap; see inside back cover.

Pelvis. The bony structure of the lower abdomen and hips, comprising the *sacrum* and three paired bones (right and left) — the ilium, ischium, and pubis — which fuse together in adults as paired *innominate bones.*

Periodontal abscess. A localized collection of pus surrounded by inflamed tissue located at the base of a tooth.

Periostitis. Infection or inflammation of the *periosteum* (the tough, vascularized membrane that covers the outer surface of bones), a condition that causes characteristic rough texture and pitting of the bone surface.

Phalanx (plural *phalanges*). Bone of the finger or toe; see inside back cover.

Phenotype. The total morphological, biochemical, and behavioral properties of an individual organism; the physical and chemical expression of the *genotype,* as also modified by the environment.

Phylogeny. The genealogy or evolutionary relationships of a group of organisms.

Pleistocene. Geological epoch of the Cenozoic Era, about 2 million years to 10,000 years ago; often characterized as the Ice Ages because of periodic southward advances of the northern glacial ice sheets; see inside front cover.

Pliocene. Geological epoch of the Cenozoic Era, about 5 to 2 million years ago, characterized by relatively dry, cool conditions compared to previous Cenozoic epochs; see inside front cover.

Plio-Pleistocene. Shorthand term to refer to the Pliocene and Pleistocene together.

Polychrome. Having multiple colors.

Polygyny. A mating system in which one male mates with several females. To be distinguished from *polyandry,* in which one female mates with several males; both are cases of *polygamy,* in which one member of one sex mates with multiple members of the other.

Polymerase chain reaction. Method of amplifying (making many copies of) a target sequence of *DNA* (a desired gene or gene segment) so that small DNA samples from a single cell can yield a quantity of identical DNA sufficiently large for a wide range of analytical techniques. Involves heating DNA to separate the two strands; binding of a "primer" molecule to the DNA strands to bracket the target sequence; and copying of the target sequence with DNA polymerase, an enzyme that links *nucleotides* together. Repeating the cycle of procedures leads to an exponential increase in the amount of the desired DNA sequence (thirty cycles produces an amplification of about 10^9!).

Polymorphism. Existence within a population of two or more *alleles* at a single *locus;* may also be used in reference to variable traits based on more than one polymorphic locus.

Positional behavior. How, when and why an animal positions itself within a particular environment, including the animal's postural and locomotor activity and the ecological relationships thereby established; an important component of how individuals (of different ages, sexes, sizes) exploit food, find mates, and avoid predators.

Postcranium. All elements of the skeleton except the skull.

Potassium feldspar. Silicate mineral containing the molecule $KAlSi_3O_8$.

Premolar. Teeth lying between the canine and molars, two in each quadrant of the mouth (upper and lower, right and left); the dentist's bicuspids.

Prosauropod. Any one of several saurischian (reptile-hipped) dinosaurs of the Triassic and Jurassic Periods (Mesozoic Era) characterized by herbivory (plant-eating) and partial *bipedality*.

Proton beam. Bombardment of a substance with energetic protons thereby causing a characteristic pattern of x-rays to be emitted that can be used to identify the elements present in the substance.

Pubis. The most ventral and anterior of the three principal bones composing the *innominate bone* (bony pelvis); or, the region overlying the pubic bone.

Punctuational. Marked by breaks or interrupted at intervals; in biology, used to denote an evolutionary event characterized by relatively sudden modification in structure over a relatively short interval of geological time; may be the result of rapid genetic evolution, or may be an apparent effect in the fossil record caused by immigration of a new population, ecological replacement, or missing geological strata containing intermediate forms.

Pyriform aperture. The pear-shaped (pyriform, sometimes piriform) nasal opening of the skull; anterior nasal aperture.

Quadrupedalism. Four-footed posture and locomotion.

Quaternary. Second (and current) Period of the Cenozoic Era, containing the Pleistocene and Holocene Epochs; see inside front cover.

Radiocarbon dating. Method that uses the known rate of radioactive decay of *carbon 14* (^{14}C) to date relatively young archeological sites ($< 100,000$ years) at which organic (carbon-bearing) material such as bone, wood, or charcoal is preserved.

Radius. One of two long bones of the forearm, the one on the lateral side as you view your forearm palm up; see inside back cover.

Reciprocal altruism. Symmetrical exchange of favors by two individuals whereby one individual temporarily sacrifices potential *fitness* in expectation of an equal return.

Recombination. The bringing together of novel combinations of genes by the process of *meiosis* and sexual reproduction.

Recursion model. Mathematical equation used to assess the change that results from a repeated sequence of effects.

Relative dating. Determining the relative age sequence of sites, artifacts, or fossils by stratigraphic position or by evolutionary or cultural sequences, without knowing the absolute age in years.

Resorb. To destruct and remove bone or parts of bone by osteoclasts (bone cells with digestive enzymes), a normal cellular process of bone remodeling and healing.

Sacrum. A bone of the *vertebral column* comprising the fused bodies of several vertebrae that serves as the point of attachment with the pelvic girdle; see inside back cover.

Scanning electron microscope. An instrument for analyzing the surfaces of tiny structures by using a focused beam of electrons to produce an enlarged image on a fluorescent screen or photographic plate.

Scapula. Shoulder blade; see inside back cover.

Sebaceous gland. A secreting organ of the skin that produces fatty or oily material; in mammals, these glands produce substances functioning as olfactory cues in social communication.

Shovel-shaped incisors. The condition of having raised lateral ridges on the back (or lingual side) of the upper front teeth (the *incisors*), as opposed to having flat, spatula-shaped teeth.

Sivapithecus. A primate genus containing several species of late Miocene apes from Asia, probably related to the living orangutan.

Sociobiology. Discipline that examines the biological basis of social behavior, especially the study of the genetic fitness of a particular social behavior in a given ecological situation.

Somatotype. Body type; physique.

Species. Among living organisms, the members in aggregate of a group of populations that potentially interbreed with each other under natural conditions; when applied to extinct organisms, a basic taxonomic category containing specimens that show a degree of morphological variability comparable to that found within living species.

Steatite. A naturally occurring massive *talc* having a grayish green or brown color that in hardness and structural integrity is ideal for carving; often called soapstone.

Sternum. Breastbone; see inside back cover.

Structuralism. A mainly historical anthropological approach to analyzing social relationships and other social phenomena by seeking abstract relational structures often expressed in a logical symbolism.

Supraorbital torus (plural *tori*). A ridge or shelf of bone above the orbits (eye sockets).

Sympatric. Of two or more populations of organisms living in the same geographic region.

Talc. A soft magnesium silicate, $Mg_3Si_4O_{10}(OH)_2$, usually whitish, greenish, or grayish in color with a soapy texture.

Taphonomy. From the Greek *taphos*, meaning burial; the study of how geological and biological factors effect the damage, decay, transport, preservation, and deposition of organic remains or artifacts.

Tarsals. Relatively small bones of the ankle and foot; in humans, the talus, calcaneus, navicular, cuboid, and three cuneiforms; see inside back cover.

Tectiform. Shaped like a roof.

Temporal bone. Complex bone of the side and base of the *cranium* that includes the inner and middle ear, the semicircular canals used in balance, the jaw joint, and a portion of the brain case.

Tertiary. First Period of the Cenozoic Era, containing the Paleocene, Eocene, Oligocene, Miocene, and Pliocene Epochs; see inside front cover.

Thermoluminescence. Method that detects the electrons trapped in ionized substances by measuring the light released from the substance on rapid heating; the ionization is created by low-level, natural environmental radiation and can be used to estimate the time elapsed since the substance was formed or last heated; compare *electron spin resonance*.

Thoracic. Of or related to the thorax, especially the *vertebrae* of the thorax that bear ribs; see inside back cover.

Type locality. The geographic place where the *type specimen* of a species was found.

Type specimen. An individual organism, or part of one (a prepared skin, a skeleton, or even just one bone) designated as the basis for erecting a formal species or subspecies name; the type specimen should demonstrate diagnostic characters that allow it to be distinguished from all other species; compare *hypodigm*.

Ulna. One of two long bones of the forearm; the one on the medial side as you view your forearm palm up, and which forms the bony point of the elbow; see inside back cover.

Upper Paleolithic. The latest (stratigraphically uppermost) division of the Old Stone Age, known for the diversity of technologies and manufactured items; usually associated with *anatomically modern humans;* see inside front cover.

Uranium-series dating. Set of methods that relies on the rate of radioactive decay of natural uranium *isotopes* (^{238}U, ^{235}U) and a thorium isotope (^{232}Th) to stable isotopes of lead; especially used in dating inorganic and organic *carbonates*.

Variant. See *cultural variant*.

Vertebra (plural, *vertebrae*). Any one of the bony segments composing the *vertebral column;* see inside back cover.

Vertebral column. The backbone, the bony unit formed of numerous individual segments, or *vertebrae,* functioning to protect the spinal cord, support the upper body weight (in animals of erect posture), and provide leverage for many muscles of the torso.

Woven bone. Bone of a coarse, fibrous texture that results from rapid, relatively disorganized growth; normally replaced later in development by more geometrically organized bone.

X-ray diffraction. Method of identifying crystalline substances or analyzing their structure by scattering x-rays (electromagnetic radiations of short wavelength) through the crystal to produce a characteristic interference or diffraction effect.

Zygote. The cell formed by fusion of two gametes; in the case of animals, the earliest fertilized cell of the embryo, formed by the fusion of egg and sperm.